Laxton's
Guide to Risk Analysis & Management

EDITED BY

Tweeds

CHARTERED QUANTITY SURVEYORS,
COST ENGINEERS, CONSTRUCTION ECONOMISTS

Guide to Risk Analysis & Management

Published by Laxton's, an imprint of Butterworth-Heinemann
Linacre House, Jordan Hill, Oxford OX2 8DP

Edited by Tweeds, Chartered Quantity Surveyors, Cost Engineers
and Construction Economists

ISBN 0 7506 2973 8

Printed and bound in Great Britain by
Biddles Ltd, Guildford and King's Lynn

Contents

Preface

This book has been written to introduce the concept and general principles of risk analysis and management and how and when they can be implemented in the construction industry. The book is aimed at construction professionals and students and attempts to demystify the subject of risk.

The use of technical jargon and complex mathematical examples has been deliberately avoided, hopefully making it easier to read and understand. Each chapter of the book is written in such a way that it should stand-alone whilst at the same time evolving into the subject of risk management as a whole. In particular, Chapter 4 Identification and Initial Assessment, introduces all the basic principles of the risk management process and in isolation can be used as a starting point for anyone wishing to adopt risk management in their organisation.

The book emphasises the need for a risk approach from the very outset and contains comments upon how risk management should be identified as an integral part of project management. Indeed, the book is written from that viewpoint. Chapter 1 Introduction to Risk Management, identifies sources of uncertainty and the need for clear and concise briefing to ensure that these areas of uncertainty are dealt with early in the project's life cycle.

The ethos behind this book it is that -if at first you don't succeed in the construction industry, you don't succeed at all! This is a very pessimistic view and it could be said that risk analysis and management is actually a pessimistic approach to the management of construction projects. However, rather than being pessimistic the aim is to be realistic which is much better than being overly optimistic. Clearly, if we can identify what might go wrong, before it does, then we can do something about it rather than waiting for it to go wrong and then being stuck with the consequences.

For ease of use the book contains a detailed glossary of terms enabling users to become more readily conversant with some of the buzzwords that risk analysts bandy around. The book also goes into some detail on some of the risk analysis techniques such as the Monte Carlo Simulation and Probabilistic Analysis and also contains a number of worked examples and case studies. The section on computers not only gives an insight into some of the software that is available off the shelf, but also contains examples of how to write simple risk analysis programmes for spreadsheets. (The examples shown are in Microsoft Excel format).

It is hoped that people will read this book to introduce themselves to risk analysis and management and to use it as a reference text and as a basis for adopting their own project specific risk management strategy.

A great deal of help has been received from a number of sources in the research and preparation of this book and I would like to acknowledge this assistance with grateful thanks. This is a first edition and has been compiled and edited by Adrian Hewitt who has received assistance from Damian Fearon and Mike Haycock. In addition David Ward and Barry Skelton from Tweeds Project Services very kindly compiled the chapter on risk assessment in connection with the CDM Regulations which I am sure will prove most useful in its own right. And finally, I would like to thank Bryan Spain who has helped in the overall preparation and presentation of the book generally.The male pronoun is used in this book. This is for ease of reading and should be taken to mean both male and female individuals.

I would welcome constructive criticism of the book, together with any suggestions for improving its scope and contents. Whilst every effort has been made to ensure the accuracy of the information given in this publication, neither Tweeds nor the publishers in any way accept liability of any kind resulting from the use made by any person of such information.

Christopher Powell
Managing Partner
Tweeds
Cavern Walks
8 Mathew Street
Liverpool L2 6RE

1
Introduction to risk management

For too long construction projects have failed to achieve the time, cost and quality targets that clients and their consultants aim for. It is widely acknowledged that there are and always will be difficulties in meeting every project objective and some degree of compromise is nearly always inevitable. Gold taps cost money, innovative construction techniques and materials can take time whilst budget constraints reduce overall quality. Time, cost and quality are pulling in opposite directions and any change in one invariably impacts on one or both of the others.

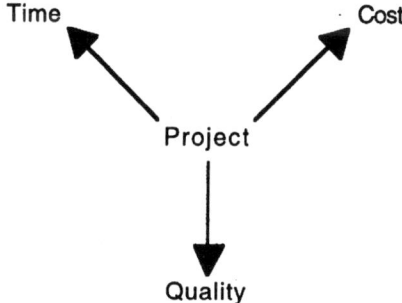

Figure 1. *Time, cost and quality*

Although each construction project is unique in nature this should not be used as an excuse for failure, we need to learn from previous experiences and put the knowledge gained to appropriate use. However, if one thing is certain in the construction industry it is that, if at first you don't succeed – you don't succeed! So what can be done to ensure success and what level of success should we be aiming for realistically. The answer to the second part of that question must surely be a better level of success than that previously achieved and if there is a raison d'étre behind risk analysis and management, that is it.

Risk analysis and management

Our aim is to ensure a greater degree of success and to minimise if not eliminate failures through the implementation of risk analysis and management (RA&M) techniques. By failures we are talking about projects that are delivered later than planned, over budget and/or to a level of quality below that desired.

Will RA&M provide us with this success? Not in isolation – it should not be seen as the saviour of the construction industry which many have suggested. But, as an integral part of project management it will benefit all those involved in projects where it is employed and without doubt, the outcome of the project itself.

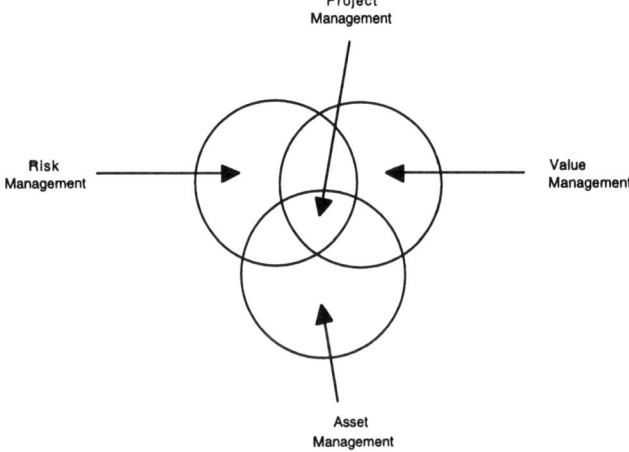

Figure 2. *Risk management as an integral element of project management*

Whilst it is recognised, indeed this book is written from such a viewpoint, that risk management is integral to project management, it should also be appreciated that it may be necessary to identify a separate, even independent risk manager. Such a person would then take overall responsibility for the production of the risk management strategy and the implementation of an appropriate regime whilst at all times remaining directly responsible to the project manager. These risk managers often act merely as facilitators to get the ball rolling in the first instance.

Briefing

The successful adoption of risk management is dependent upon the stage of the project at which it is introduced. Much of the ethics behind risk management is the identification of risks before they materialise, followed by the implementation of mitigation strategies and contingency plans so that if and when they do materialise their potential impact is reduced. Clearly, if risk analysis and management techniques are not put into use until late in a project's development then their effectiveness in ensuring project success is greatly diminished. It is essential that at the project briefing stage consideration is given to a risk analysis and management strategy. Indeed, risk identification described in more detail later in this book, has been found to be of great benefit in developing the overall project brief.

It is widely acknowledged that deficiencies in the brief are often the source of the uncertainties that are the cause of the risks. Early consideration of risks helps the project team to appreciate that:

- their building is unique and likely to have its own specific problems
- buildings involve major capital and longer term revenue expenditure over which most clients need to keep a tight control
- by comparison to other expenditure, building projects can take a long time and may involve much abortive work

- a large number of people with different viewpoints are often involved and conflicts are not unusual – it is important that such issues are dealt with early in a project's life.

Risk analysis and management must help towards the development of the brief by making project team members consider in detail what are the pitfalls of their development and then assess what action can be taken to alleviate or avoid them.

Risk efficiency

Examination of these potential pitfalls along with the project's overall time, cost and quality targets, enables us to consider risk efficiency. Risk efficient projects are those delivered at the lowest cost, compatible with the specified quality, in the desired time and with the minimum risk exposure.

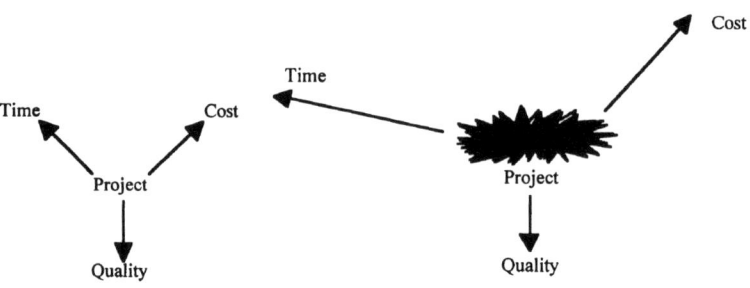

Figure 3. *Uncontrolled risk exposure*

Risk exposure can be looked at like an undetonated bomb – it may explode, it may not. If it does explode it could wreak havoc but if appropriate contingency plans are put in place, its impact can be controlled. Risk management facilitates this controlled explosion.

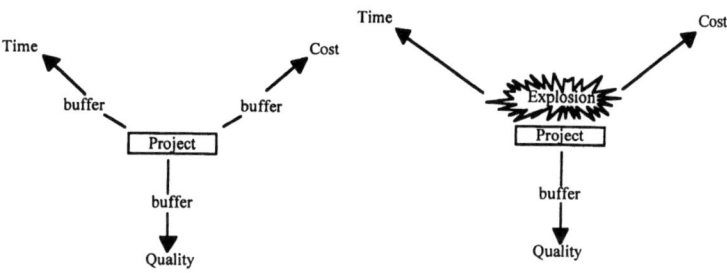

Figure 4. *Controlled risk exposure*

Good luck or good management
It is this use of risk management to install contingency plans (buffers) thereby ensuring risk efficiency, that should be considered good management. The bomb failing to explode is merely good luck!

However, it must be appreciated that simply increasing the buffer zone to ensure that time, cost and quality targets will always be met is not an appropriate approach. It is often necessary to operate within constraints and therefore oversized contingencies of any form will not be acceptable. Indeed such an approach may result in the project failing to get off the ground altogether.

This introduction has attempted to outline why there is a requirement for risk analysis and management and why it should be considered an integral part of project management. The remainder of this book will deal with the implementation of a project specific risk management strategy and the techniques involved in risk analysis and management. The intention here is to describe methodologies that are widely acknowledged and accepted.

What is risk analysis and management?
This book is intended to introduce the general concept and principles of risk analysis and management that should be implemented as an integral and essential aspect of project

management through a detailed and dedicated project specific risk management strategy. This book goes on to describe the broad principles necessary to produce and enforce an appropriate risk management regime. Whilst it should be appreciated that these broad principles will apply to all schemes, the extent to which the techniques are implemented will be very much dependent upon the nature, complexity and size of the construction project being undertaken.

Whenever you read articles or facts about risk analysis or risk management ,the words uncertainty, risk and hazard appear. Often they are not distinguished between and to some degree, the same will be the case for this book.

Brief definitions taken from the *Oxford Dictionary of Current English* are:

- uncertain – not certainly knowing or known; not to be depended on, changeable
- risk – chance or possibility of loss or bad consequence
- hazard – source of danger or risk.

We can therefore conclude that risks generally arise because either there is a lack of certainty about the project being undertaken or that hazards exist within the project. Broadly speaking, hazards and uncertainty are considered the main sources of risk. If we can identify what hazards exist, the circumstances that may arise and the areas of uncertainty that lie within the project, we can begin to assess the level of risk to which we are exposed. This level of risk exposure will then influence the decision-making process and ultimately how and if the project will proceed.

Uncertainty
   &   =>   *Potential Risk Exposure*
Hazard

So when will these uncertainties and hazards exist and how will they materialise into risks? Later chapters will deal with this in more detail along with the action that needs to be taken when this happens. However, it is clear that:

- hazards always exist but the degree of risk exposure is usually a direct result of action taken, e.g. existing underground services crossing a site do not become a risk until you start digging in the ground
- uncertainty exists in most instances due to a lack of project definition.

Project definition

An important key to the development of any construction project is the briefing given to the project team by the client or his representative. One of the greatest sources of uncertainty is the production of this brief, i.e. the client clearly defining what is required. Only the client can decide what he wants, when he wants it and how much he is prepared to pay for it.

A lack of a detailed brief nearly always results in compromise later in the project's life cycle. The life cycle of a building can be generally defined as follows:

- inception and briefing
- feasibility
- design
- construction
- occupation
- obsoletion and subsequent demolition.

The client is primarily concerned with the finished article and the occupation of the building. He is the end-user. If the brief contains inadequate detail he will not get what is required and may incur additional expense in rectifying the situation. The stage at which this rectification takes place has a significant impact upon the overall cost of the project. Rectifying a problem during the design phase will result in abortive time by consultants. Rectification during construction will worsen the impact but if left until occupation, it can be extremely costly and time-consuming. Getting it right from the start with a clearly-defined project brief is extremely important. Trying to arrive at a final design without making an assessment of the client's or end-user's requirements will prove extremely costly.

A project brief should contain:

- details of the project participants along with a definition of their roles, extent of involvement and constraints
- who is the client's representative and responsible for delivering the project to the client on time, within budget and to the desired quality
- who are the end-users and what is their involvement and do they have a say in the final outcome
- what third parties are involved and what are their responsibilities and constraints
- financial issues – how is the scheme to be funded, are there any grants or subsidies involved and what is the agreed budget
- legal restrictions or requirements
- project's life cycle including target dates for completion and occupation
- the current status of the project, what decisions have already been made and can they be rescinded
- site-specific issues such as location, current use, past use and characteristics, i.e. size, topography and ground conditions
- use and building activity such as occupancy levels and characteristics of occupants
- quality of finished product required.

As stated previously many conflicts exist or are caused by poor briefing or changes in the brief. Poor briefing leads to uncertainty but by adopting a risk management strategy, areas of uncertainty can be identified. Once the uncertainty is identified, it can be managed with confidence to ensure that the consequences of the changes that will take place, due to those uncertainties, can be controlled and maintained within the project's constraints. Appropriate RM facilitates controlled change as opposed to uncontrolled change.

To close this chapter it is considered worthwhile to differentiate between risk analysis and risk management. These views remain consistent throughout the book.

Risk analysis

This primarily involves the two stages of:

- identification/qualitative analysis the formal identification of risks and an initial assessment of their likelihood and potential impact
- assessment/quantitative analysis the detailed quantitative assessment is the area most people identify as risk analysis. It often involves quite sophisticated statistical calculations in an attempt to accurately predict the potential impact of risks should they materialise.

Risk management

This book views RM as the whole, i.e. the identification and assessment of risk that is known as risk analysis plus the adoption of appropriate responses and mitigation strategies. Furthermore, RM is an integral part of project management. It should encompass a regime that will formally govern the allocation of both individual and overall risk responsibility and the methods of reporting that provide the evidence that appropriate risk management is taking place. Put together with the aim of ensuring that the project is delivered within its time, cost and quality parameters and you arrive at a RM strategy.

Summary

To summarise, this chapter has:

- introduced the concept of risk management as an integral and essential element of project management
- defined risk efficiency as the delivery of projects at the lowest cost, compatible with the specified quality, in the desired time and with the minimum exposure to risk
- highlighted that RM aims to deliver projects through good management and not good luck

- explained the differences between hazards, risks and uncertainties
- emphasised the need for a clear and well-defined project brief
- identified the need for an overall risk management strategy.

This RM strategy is the subject of Chapter 2 of this book.

2
The risk management strategy

Introduction

A risk management strategy can be considered as a framework
for managing risk to ensure that time, cost and quality targets
are met. As stated previously it is an integral element of
project management – indeed RM is seen by many as the
raison d'étre of project management and has led to the increasing
use of the term risk driven project management. The adoption
of a RM strategy is an attempt to approach projects from an
analytical and common-sense viewpoint through a system which
can be as narrow or as wide-reaching as deemed appropriate.
Therefore, at the inception of a project when briefing discussions
begin, the outline of the project specific risk management
strategy should also be considered. The basic elements of the
RM strategy are:

- the processes that are to be adopted for identifying
  and assessing risk and methods of response and
  mitigation
- the regime that governs how information arising from
  the risk management process is administered
- a definition of the relationship between the RM strategy
  and the overall project management aims and objectives
  of time, cost and quality.

Risk management and the project manager

The term project manager means different things to different people. It could be described as that person or body responsible for the overall delivery of the project to the time, cost and quality targets required by the client or end-user. Whatever the definition, the problems are nearly always the same:

- the brief is very often inadequate and a poor reflection of actual requirements
- the project manager has to put in place a management structure to control a multitude of different people and disciplines, many of whom he has no experience of and may have their own individual thoughts, aims and objectives
- the construction project is bespoke, never having been undertaken before and therefore past experiences are limited.

However when projects fail clients may state... well I told you what I wanted... whilst project managers will argue... look what I had to contend with... Realism lies somewhere between the two.

In the opening chapter it was stated that risk management is not a panacea, but, through the adoption of a RM strategy within the overall control of the project manager the degree of confidence about the project outcome should be increased. If the project manager is seen as having the responsibility of the overall delivery of the project he should also have overall management responsibility for the risk inherent in that project.

Risk management is seen by most as comprising identification, analysis and response. This book takes the view that this is simply the risk management process and that RM as a whole will cover wider reporting and responsibility issues which ultimately makes RM an integral element of project management.

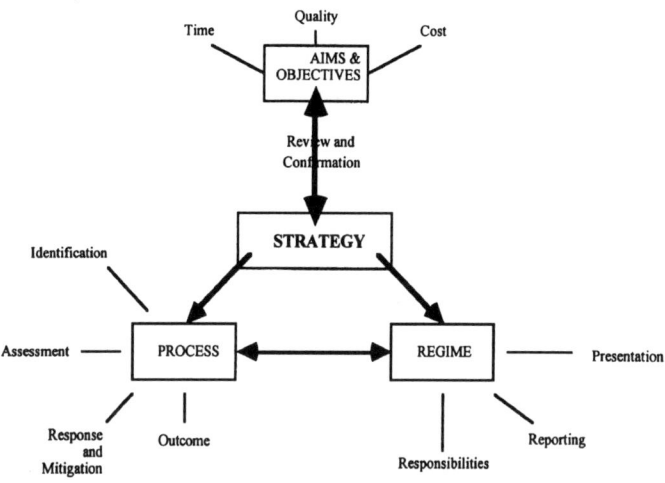

Figure 5. *The risk management strategy*

The RM strategy should define the process which will be utilised within a regime, all of which is implemented to achieve the overall aims and objectives of time, cost and quality.

The process

The RM process is discussed in detail in Chapter 3 of this book but essentially comprises three basic steps:

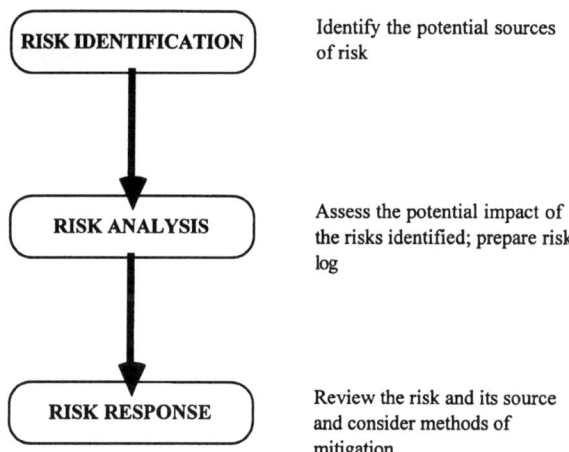

Figure 6. *The basic risk management process*

However, although these are the three basic steps of the risk management process, clearly it cannot be totally sequential in nature. Risks are likely to arise out of the action taken to respond to initially identified risks – these are known as secondary or resultant risks and are dealt with later in this book when looking specifically at the RM process.

The regime

The RM regime is also discussed in more detail later in this book but will broadly consist of:

- overall RM responsibility
- allocation of individual responsibility
- necessary and appropriate reporting procedures
- management of contingency through controlled change.

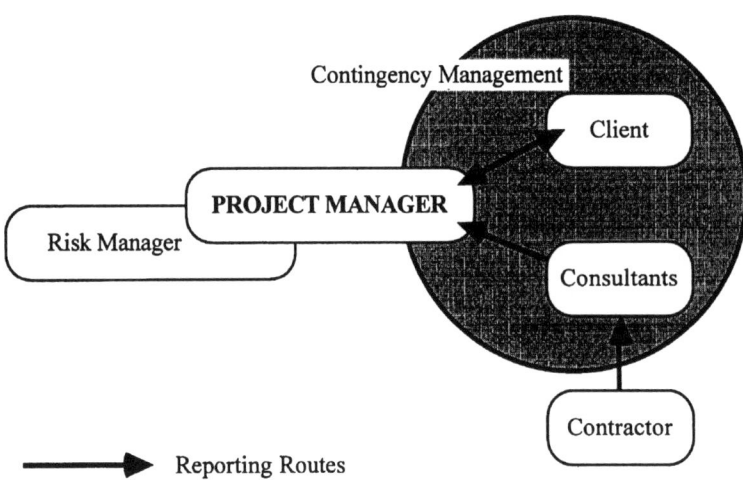

Figure 7. *Typical RM regime*

Effective RM is reliant upon the formulation and implementation of a pre-determined regime that will govern reporting, presentation and responsibilities. It may be necessary for the regime to be prescriptive and explicit either on a project basis or at client level, to ensure a degree of commonality. It is essential when reporting both up and down the hierarchical tree that consistency exists and comparisons with previous reports/ projects are on an apples with apples basis.

The continual monitoring, review and re-assessment of risk will only result in effective management and control if carried out in conjunction with appropriate, concise and accurate reporting procedures. Therefore, great emphasis should be placed upon adherence to the reporting procedures proposed in the RM strategy. The responsibility for the formulation and implementation of an appropriate methodical and workable reporting framework lies firmly with the project/ risk manager.

As a minimum, reporting should be undertaken on a monthly basis. However, it is recognised that, dependent upon the specific project, its nature and stage of development, it may be necessary for additional unscheduled interim reports to be made. Indeed, it is a requirement that upon the realisation of any changes in circumstances that materialises into a risk or potential risk with a significant impact upon the project outcome, it should be reported on immediately, irrespective of any pre-determined reporting or deadline dates. Nevertheless, it is essential that the project/risk manager implements a co-ordinated procedure for reporting, ideally based upon a suite of compatible pro-formas. He must liaise fully with other members of the project team and ensure that they are aware of what information is required, in what format and by what time.

It must be stressed that this applies to all participants whether above or below in the project team hierarchy. The implementation of this predetermined procedure for reporting is the main tangible evidence that the management of risk is taking place and provides the facility for carrying out a project risk audit. Quite simply, compliance with an appropriate reporting procedure will enable the audit trail of any risk to be examined through all the stages of identification, assessment, response and mitigation.

A comprehensive set of reporting procedures should also enable projects to be analysed in detail thus allowing records of past experiences to be obtained which should improve the general level of corporate knowledge. It will also allow for the audit of the regime itself and adherence thereto.

In terms of defining responsibilities within the RM strategy, the aim is to formulate procedures that will define the flow of information across and between project participants. This must be undertaken whilst, at the same time, appreciating the various degrees of confidentiality that may exist.

Summary
This chapter has briefly examined the need for a project specific or client specific risk management strategy to ensure that risk is processed in a consistent manner and reported upon through a regime employed to avoid confusion. By adopting such a prescriptive approach clients can proceed with their projects with confidence. However, it should be appreciated that the aim is consistency and not the enforcement of a series of procedures that will be used like the proverbial big stick by over-zealous clients and their representatives. Achieving success through good management and having the confidence of the people around you is the aim rather than relying upon good luck and being fearful of the consequences.

It is impossible to identify all the potential pitfalls all of the time, that is why the RM process is an ongoing iterative one.

3
The process for managing risk

Introduction

This chapter aims to introduce the concept of risk management as an ongoing, iterative (i.e dependent upon the results of itself) process. Whilst every project undertaken is undoubtedly different, clearly the approach to the management of risk can be similar, as, in most instances, the key activities exist. Namely:

- identifying the risk
- assessing its potential impact
- responding to it
- taking an appropriate course of action to avoid or mitigate the potential impact of the risk.

This is best explained in Figure 8 below:

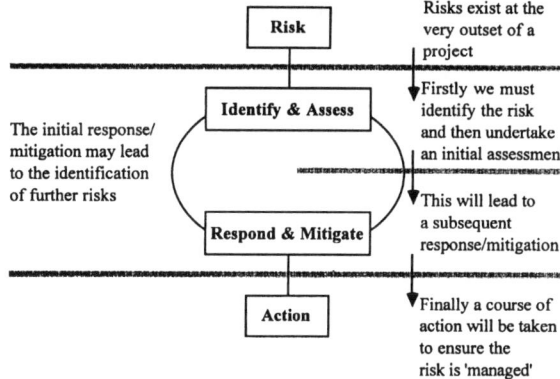

Figure 8. *The risk management process*

To further emphasise the cyclical nature of the RM process this can be enhanced to highlight the results and impact of the actions taken.

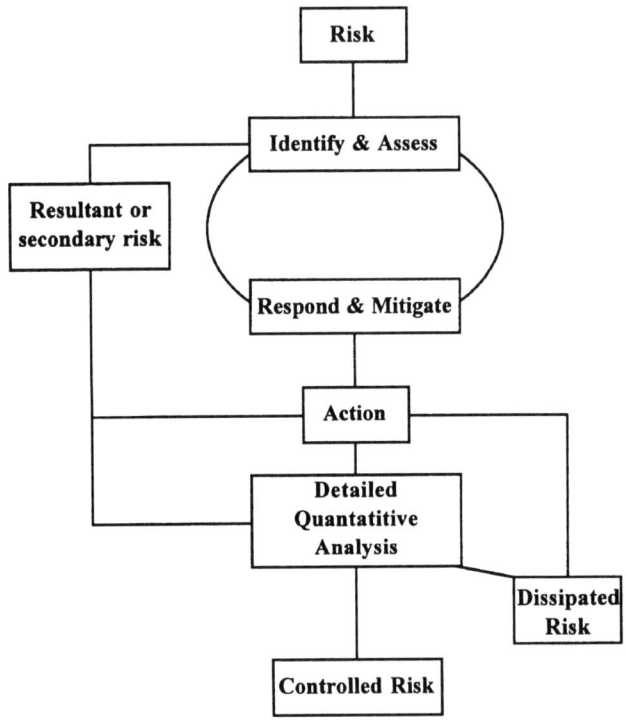

Figure 9. *Detailed risk management process*

The various elements of this RM process are outlined on the following pages and explained in detail later in this chapter.

Identification

As stated previously, risks generally arise as a result of uncertainties. Where uncertainty surrounds a project there is likely to be a distinct lack of confidence in the potential outcome. However, if we can identify where this uncertainty exists we can go a long way to managing that uncertainty and therefore the resultant risks that are likely to manifest themselves.

All projects involve risk some of which will be obvious, others much less so. Before we can quantify their likely effect and before the risk can be managed, it must first be identified. Methods for identifying risks could include:

- brainstorming
- interviews
- questionnaires
- use of specialists
- previous experience.

Formal risk identification procedures are discussed in more detail later in this chapter. In identifying risks, it should be appreciated that such an exercise can never be undertaken too late. If Risk Management is seen as part of the overall decision-making process and the identification of risks is perhaps the most important element of risk management, then this action must be taken early in a project's life. It is well known that many of the major decisions that have the greatest impact on the project are made during its early feasibility and design development stages.

It is at this stage that changes can be made with the least disruption. But it is also at this stage that the information upon which such decisions are made, is most likely to be incomplete or inaccurate. Therefore to ensure that the right decisions are made, all the important risks must be identified and assessed at the earliest possible point in the project's life cycle.

Assessment

The purpose of the risk assessment is to understand and quantify the likely effect of any potential risks. Essentially this assessment phase is split into two:

Qualitative assessment   –   where the source, cause and effect of the potential risk is reviewed and described in detail. By doing this, a risk register can be compiled where the status of each risk can be considered and updated on a regular basis. This is at the core of on-going RM.

Quantitative assessment   –   here the likely effect of the risk is analysed in detail along with the resultant knock-on effect on the overall outcome of the project. By undertaking such detailed analysis the most likely and worst case project outcome can be calculated.

The cyclical and iterative nature of the RM process is highlighted by the fact that often after an initial qualitative and quantitative analysis, additional risks may be identified. The detail of qualitative and quantitative assessment is dealt with later in this chapter.

Initial response and mitigation

Initially, it is often not possible or indeed appropriate to separate the activities of risk identification and qualitative assessment and considering a necessary response/mitigation. Again, this is a result of the cyclical nature of the RM process. Risk management helps to control risk but to realise the full rewards it must be appreciated that it is a management task that needs to be undertaken. It is essential to successful risk management that risks are managed through considering appropriate courses of action that respond to risks and mitigate their potential impact.

Responses to risk may include:

- risk avoidance
- risk reduction
- risk transfer
- risk sharing
- risk retention.

In considering response/mitigation to risk, thought should

be given to the cost of the course of action proposed. In some instances it may be more appropriate to retain a risk and suffer the consequences should it materialise, e.g. it is a general policy that the Crown do not insure because the cost of insurance is likely to outweigh the cost of paying out claims/damages when they arise.

Furthermore, risks need to be allocated to those parties best placed to influence both the likelihood of the risk occurring and its potential impact should it occur.

The responsibility for the allocation of risks will lie with the project manager or his designated risk manager. Prior to the allocation of risks to the various project participants it is necessary to consider a number of factors:

- who is best placed to influence the events that may lead to the risk materialising
- who is best placed to control and manage the risk when it does materialise
- is the risk one which lies within the project's control
- the timing of risk allocation.

It should be appreciated that the allocation of a risk to the wrong person for the wrong reasons and at the wrong time can have almost as adverse an outcome as not identifying and responding to the risk in the first place.

Action

By this we are simply identifying a course of action that is a result of the response and mitigation strategies considered. For example, having identified a particular design risk, the recommended response may be to reduce the risk and this could be executed by undertaking a re-design. This re-design is considered as the action resulting from the response and mitigation strategy. The action taken could result in a number of possible outcomes:

- resultant or secondary risk clearly these are additional risks which are identified as a result of the action taken. Once identified, these resultant risks are subject to the same assessments, response/mitigation approach that make up the RM process

- dissipated risk these are risks which, through the action taken, no longer have any significant effect on the project
- residual or retained risk these are the risks which remain within the environ of the project and are subject to the detailed quantitative analysis. It is these risks which, when quantified make up our contingency and are continually monitored and controlled throughout the project's life cycle.

Summary

This chapter has outlined briefly the overall risk management process and highlighted that this process is one of an iterative and cyclical nature. The following chapters will look at the various elements of the RM process in more detail.

4
Identification and initial assessment

Introduction

The construction industry can be subject to an exceptionally broad range of risk and uncertainty. Procuring a building from inception to commissioning may involve a great number of people all with vastly different specialist skills and responsibilities. Each new project entails unique design and construction problems because most construction projects are bespoke.

Without first identifying what the potential risks are we cannot ascertain:

- if they will arise
- what effect they might have if they do arise, and
- what measures need to be taken to prevent their occurrence.

Clearly the identification of risks may be considered the most important stage in the risk management process. When identifying risks, previous experience should be considered an invaluable asset. All project team members should be encouraged to share any relevant knowledge gained from previous circumstances of a similar nature.

But the use of historical data and previous experience has its limitations. Although the knowledge gained from historical projects may assist in the prediction of future problems, the uniqueness of each project may present circumstances not previously encountered. Furthermore, it should be appreciated that it may be restrictive in that those involved do not always consider the wider aspects of the projects and may solely concentrate upon their own previous experience.

Risks stem from uncertainty which is mainly caused by a lack of detailed information at the time a decision is made. Uncertainty can be defined as a situation about which there is no historical data or previous experience. An example could be a new building which utilises an innovative construction material which has not been utilised before. Therefore there would be no historical data on which to make fundamental decisions on methods of working.

It is important to differentiate between the sources of risk and their effects. Ultimately, most of the risks found on a construction project will have an adverse consequence on one or more of the following attributes of a project:

- TIME   –   additional time for design, construction and subsequently the occupation of the building
- COST   –   additional costs
- QUALITY –   failure to meet required quality.

It can be seen that the effect of risk events are generally a form of loss to the client or building owner. Consultants on a building project have a professional duty of care to their clients and in the present economic/legal climate clients have shown an inclination towards legal action when they have suffered loss as a direct consequence of poor advice.

Therefore it is of major importance to identify the source, trigger events and effects of risks to attempt to alleviate any consequence of a risk occurring.

Controllable and uncontrollable risks

Risks arise as a result of actions or events that are either within or outside the project's control. Thus, they are termed controllable and uncontrollable risks.

Controllable risk   –   is a risk which is within the control of the project or can be controlled by the project participants, e.g. the price of the specified facing bricks increases. The outcome of this occurrence is the project team can explore alternative (cheaper!) facing bricks. The situation is within the project team control.

Whereas an uncontrollable risk is a factor which is outside the control of the project team and cannot be influenced in any way. Such risks very often originate from external sources, environment – political, economic or climatic.

Uncontrollable risk   –   there may be a new requirement of the fire officer which must be incorporated into the fabric of the building before a fire certificate can be granted. The project has no choice but to conform.

There are basically three types of risk:

- factors within your control
- factors in the control of others, e.g. Planning Requirements, Building Regulations, Government taxation, Banks – rate of interest
- acts of God – outside your control, e.g. weather.

The specific type of risk encountered and who, if anyone has control over it, will decide both the nature of the response to it and also who is allocated the responsibility for managing that risk.

Risk identification

It is generally accepted that construction projects are unique in their nature and each project will attract different risks. Whilst the knowledge gained from the experience of previous projects (particularly failures) may assist in the prediction of future problems, it is unlikely that the same particular combination of risks will be encountered. It is therefore essential that new projects are treated on their individual merits.

Risks exist from the very outset of a project. Therefore we need to identify what they are, ascertain when they might arise, what effect they may have and what measures need to be taken to prevent their occurrence or mitigate their potential impact. The identification of risks may be considered as the most important stage in the RM process, if only in terms of bringing considerable benefit to all parties in the greater understanding of the project, irrespective of whether further action is taken or not.

Previous experience should be considered an invaluable asset to any identification and consultative stage. The use of previous experience will manifest itself across the whole spectrum of identification techniques, in that all project team members should be encouraged by the project manager to draw upon and share any relevant knowledge gained from previous circumstances of a similar nature. However, construction projects will present a unique set of circumstances and therefore, previous experience alone is unlikely to generate the level of understanding necessary to fully deal with the circumstances which are particular to the current project.

Drawing upon previous experience can significantly foreshorten the risk identification process. But it should be appreciated that it may be restrictive in that those involved do not always consider the wider aspects of the project and may solely concentrate upon their own field of expertise, i.e. a blinkered approach, simply utilising a schedule of identified risks from a previous project. Also, due to the time lapses between projects it may also be the case that what is considered previous experience is an inaccurate reflection of actual events thereby giving rise to misleading information.

When identifying risks it is important to appreciate not merely the risk itself but the source, the event that may lead to the risk materialising and the effect of the risk if it does materialise. It is, therefore, of great importance that when identifying risks, generic terms are avoided and that the risk is described in detail, e.g.:

underground services   this gives very little information about
                        the risk itself and is almost impossible

to mitigate

9" high pressure water pipe crossing site in north west corner at depth of 1.5m below existing ground level — obviously gives more detail and enables the project team to consider a course of action to mitigate the risk.

Typical sources of risk may be:

SOURCE	EVENT	EFFECT
Scaffolding incorrectly assembled	Scaffold collapse	Injury/Death to workmen
Inflation	Inflation rises above that included in estimate/tender	Estimate/ tender becomes inaccurate
Weather	Exceptionally inclement weather	Project is delayed

Figure 10. *Sources of risk*

If such importance is placed upon identifying the source of risk not just generic headings or terms clearly the method of risk identification must be given some thought. There are numerous methods for identifying risk and include:

- brainstorming
- interviews
- previous experience
- questionnaires.

Brainstorming

This is a very effective procedure for the identification of risks and the subsequent response/mitigation thereof. Brainstorming basically involves open, frank and in-depth discussion in order to attempt to impart from the project participants:

- what peoples' concerns are
- where are the areas of uncertainty/hazards
- what risks are likely to arise
- likelihood of occurrence of any risks
- potential impact of the risk
- project participants' initial response to the identified risks.

Brainstorming sessions take the format of an informal meeting with no set agenda and no definite timescale. It is of paramount importance that an experienced chairman (usually the project or risk manager) is appointed to facilitate a meaningful discussion and to document the proceedings. This normally takes the form of a risk log or an individual risk identification form for each risk identified

These informal brainstorming sessions should normally be attended by the client, project manager, risk manager and if applicable, appropriate members of the design team and the end-users of the development.

Furthermore, depending upon the procurement route and the contractual relationships, it may be beneficial to include in these sessions the main contractor(s) and any specialist suppliers/sub-contractors. However, different procurement routes may not easily fit into this sort of management technique. If a tender is procured in competition, many companies believe that by identifying risks at an early stage and including cost contingencies for them, may result in pricing themselves out of the job. Therefore any risks the contractor may identify could become the basis of a financial claim at a later date.

Consideration must be given to the numbers who are attending and an upper limit may be necessary. But it is generally advisable to establish this exercise as a single activity in order to maintain cross-fertilisation of ideas and prevent the development of opposing factions. If it becomes apparent that the project

Project _____

Risk Reference. _____   Date Identified _____

Brief Description of Risk   _____

_____

_____

Identified By   _____

Likelihood of occurrence   *low/medium/high**   or _____ %

Potential Impact   *low/medium/high**   Impact area   *time/cost/quality**

Risk Exposure   *acceptable/insignificant/significant/critical/unacceptable**

Initial Response / mitigation   *ignore/manage/design/share/transfer**

Brief Assessment of Impact

Likely Time   _____   Items affected   _____
Worst Case   _____   _____
                                                _____

Likely Cost   _____   Description   _____
Worst Case   _____

Likely effect   _____
on Quality   _____
                    _____

Secondary Risks   *yes/no**   Description   _____

Reference of   _____
secondary risk   _____   _____

Responsibilty   _____

Detailed assessment required?   *yes/no**   By _____

Signed   _____   Verified _____
For   _____   For _____

Figure 11. *Typical risk identification form*

Ref	Description	Probability	Potential Impact	Response/ Mitigation

Figure 12. *Typical risk log*

participants are too numerous to sustain workable discussion or it is becoming too expensive an exercise, then alternative methods of risk-identification should be considered. Also this method is susceptible to being dominated by stronger personalities and good skill is required from the chairman or facilitator to ensure the meeting is not hijacked. In some instance the opposite may be the case and it may be necessary to promote discussion through the use of checklists, e.g.:

Third party risks	planning requirements
	legal agreements
	covenants
	environmental issues
	pressure groups.
Site risks	access restrictions/limitations

	existing occupants & users
	existing buildings
	boundaries
	security
	land purchase & ransom strips
	existing services
	ground conditions.
Client risks	briefing
	timescales
	financial issues
	quality standards.
Design risks	interpretation of brief
	design development
	design demarcation
	timescales
	estimating
	professional ability.
Construction risks	procurement route
	programme
	variations
	site management
	liquidation/insolvency
	latent defects
	health and safety.
External risks†	market conditions
	political conditions
	government legislation
	weather.

† this source of risk would generally be considered uncontrollable whereas the other sources are all within the overall control of the project team.

Interviews

This is a technique that has been used historically by personnel departments and other consultants to extract information. It has also been used by risk managers to identify possible risks in a development.

Interviews allow relaxed yet meaningful discussions to take place in a relatively formal manner. Ordinarily, interviews will be co-ordinated and undertaken by the project or risk manager, who will be responsible for providing documentary records of each individual interview and collating the results gathered. The format of the interview should be prepared in advance, pre-prepared questionnaires and discussion material may be issued to the interviewees prior to the interview date.

As with brainstorming, interviewees should include the full project team. The aim is to facilitate a discussion which will reveal the individuals risks, doubts and reservations about the project. Questionnaires can be used by the manager to attempt to stimulate areas of discussion which may not have been covered.

The interviews may take place on a one-to-one basis or on a many-to-one basis. The many-to-one basis should consist of project members from different disciplines so that subjects raised can be viewed from different perspectives. The problem with this method is that it is time-consuming not only to carry out the interviewing but to also record (and analyse) the risks arrived at therefrom.

The Delphi technique

The Delphi technique attempts to produce objective results from subjective discussions. This method may be applicable to the identification of risks but it is more suited to attaching likelihood of occurrence and potential impacts of previously identified risk events. This method basically involves the following sequence of events:

- a questionnaire is forwarded to all the appropriate members of the project team by the appointed risk or project manager
- the members of the project team give their objective views in response to the questionnaire and return them to the risk manager
- the risk manager then collates these results and redistributes them. Each project participant now receives

a different set of views and is requested to reconsider their original answers and resubmit them to the risk manager
- these revised results are again collated by the risk manager and redistributed again in the same manner as above
- this iterative process is continued until the risk manager is satisfied that a consensus of opinion has been reached.

The main advantage of this method is that all participants act independently, there are no stronger personalities to dominate nor any peer pressure. The main problem is that this is a very repetitive technique and is therefore time-consuming. Also, while a consensus opinion may be reached, that often means an opinion that no one person offered in the first instance.

Expert systems

Expert systems are developed using historical data and experience aided by specialists in each discipline of the construction team to attempt to identify possible risks. They tend to be expensive to produce particularly in the construction industry due to its bespoke nature.

Expert systems may not reveal any hidden risks as they tend to concentrate on the obvious risks which have come about previously on other historical projects or upon the area of expertise of the specialist who put the package together.

Expert systems are currently improving and development continues especially using computer software packages. There is very little evidence that suggests expert systems will identify all the risks inherent in a construction project but will only show standard risks. These models should be used with great care.

Questionnaires

Questionnaires are usually drawn up from a combination of previous experience and specific project criteria. There are two forms of questionnaire, one is a very general form with non-specific prompts or questions – the other can be as detailed as is

RISK QUESTIONNAIRE

Job Title:

Client:

Risk Category:

Sheet: ☐ of ☐

Risk	Impact	Probability

NR.

     H M L
T ☐ ☐ ☐   ☐ H
C ☐ ☐ ☐   ☐ M
Q ☐ ☐ ☐   ☐ L

NR.

     H M L
T ☐ ☐ ☐   ☐ H
C ☐ ☐ ☐   ☐ M
Q ☐ ☐ ☐   ☐ L

NR.

     H M L
T ☐ ☐ ☐   ☐ H
C ☐ ☐ ☐   ☐ M
Q ☐ ☐ ☐   ☐ L

NR.

     H M L
T ☐ ☐ ☐   ☐ H
C ☐ ☐ ☐   ☐ M
Q ☐ ☐ ☐   ☐ L

General Comments/Notes

Organisation [   ]   Date [   ]

Figure 13. *Typical risk questionnaire*

required by the particular project.

Before being distributed, questionnaires are compiled by the project or risk manager in a methodical and structured manner ensuring that areas of potential risk are exposed for examination. Careful consideration should be given to the timing of the issue and subsequent return of the forms so that it fits into the risk management process and in particular is used in conjunction with other identification techniques.

It is recommended that questionnaires are not used in isolation but are issued to a sufficient number of the project team as a precursor to, and/or in conjunction with, other identification techniques. The questions may act as a stimulus and trigger other thought-processes. These forms when utilised will allow for more open and frank disclosure of risk without dominance from stronger personalities and no peer pressures.

Questionnaires also facilitate consistently presented answers from the different team members which allow less time-consuming and more meaningful comparisons. Therefore the risk manager can ascertain more readily any apparent consensus.

The main disadvantages of questionnaires are that the compilation of the questions may originate from the risk managers own preconceived ideas. Therefore considered and possibly lateral thought will not be promoted amongst other team members.

Combined approach

In reality it is recommended that an approach combining a number of identification techniques is adopted – methods of identification are not necessarily limited to those mentioned in this chapter. By combining techniques the various disadvantages of one method may be offset by the advantages of others to produce a more meaningful exercise

Throughout the whole stage of the identification process it is of paramount importance that all information disclosed by all project members is treated in a confidential manner. However, it is recognised that certain circumstances may lead to non-disclosure of information because they may be of a strategic or even political nature.

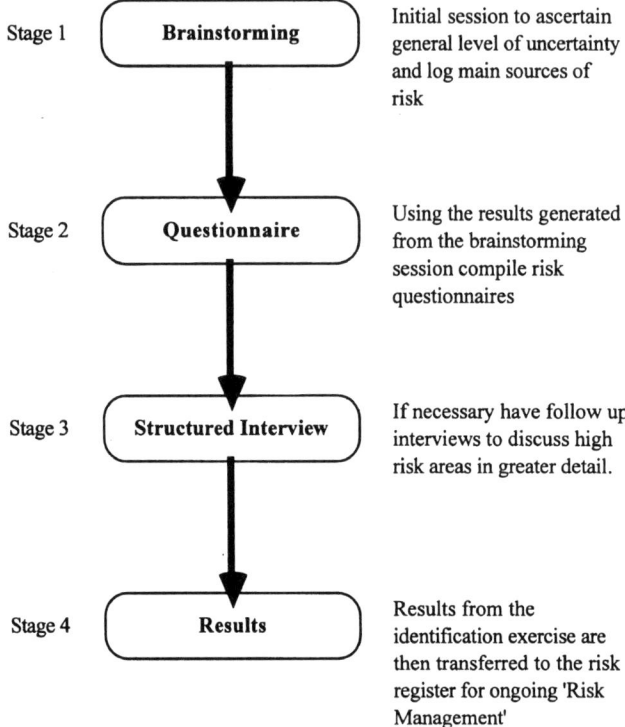

Figure 14. *Combined risk identification*

Ultimately this identification stage should, following vigorous examination, produce a schedule of risks derived from a consensus of opinion of the project member. However, it should be appreciated that risks could be identified throughout the project's life and in order to ensure that these risks are included in the risk management process, a method of recording new risks must be derived. The risk identification form indicated in Figure 11 is an ideal method of logging new risks. Having identified the risks, they should subsequently be assessed. The assessment stage generally falls into two distinct phases:

- qualitative assessment
- quantitative assessment

Initially we will concentrate on the qualitative assessment.

Qualitative assessment

This stage involves the registration of the identified risks in a formal manner. An acceptable method of formalising this process is to utilise a qualitative assessment proforma (see sample forms) accompanied by a risk register which may include nor necessarily be limited to the following:

- classification and reference
- description of the risk
- relationship of risk to other risks (interdependencies)
- potential impact
- likelihood of occurrence (probability)
- response/mitigation strategy
- risk responsibility/owner.

The risk management process is iterative and cyclical by nature and it is likely that, even at this initial qualitative assessment stage, some idea of what should be included under the headings on the risk register will be known.

The nature of the process ensures that information will be revised and, if necessary, updated some time later in the project's life cycle. Where this is the case, and in order to differentiate between initial intuitive information and that of a more considered nature, it may be prudent to adopt some form of keyed annotation that clearly indicates the status of the information.

Classification and reference

- risk should be classified into categories which reflect that area or stage from which uncertainty and therefore subsequent risk has arisen
- classification is an aid to identifying the source of risk and could include the following:
  environmental – site conditions, health and safety
  contractual   – client, contractor, third party (sub-contractor), information protection.

> *design*   –   planning permission, fire officer's
>   requirements.

- referencing of the risks can be based upon a unique reference, allocated dependant upon its classification (by adopting such an approach, a recognisable audit trail can be established)
- to further refine the source of risk it may be useful to consider the stage of the design as defined in the RIBA Plan of Work or indeed to use broader risk management stages as defined below.

RM STAGES	WORK STAGES	OCCURRENT RISK CLASSES
a) Project identification and strategy	Project identification Project strategy Appointment of consultants	Site Client Design Third party Other
b) Design definition	Project planning/ Preliminary design	Site Client Design Third party Other
c) Detailed design and construction	Detailed design; Tender preparation Final project approval; Tender invitation; Evaluation and contract award	Site Client Design Third party Contractor Other
	7. Construction period	Site Client Design Third party Contractor Other
d) Occuption operation and demolition	Maintenance period and completion of account Long term Maintenance and functional operation Decommissioning	Client Third party Contractor Other
Those risk classes in bold indicate what may be considered the most likely sources of risk at the particular stage identified		

Figure 15. *Plan of work*

Description of the risk
This is quite simply, as the title implies, a brief description of the risk. However, this description must be unique in order to avoid confusion with other similar risks in the RM process.

Risk trigger
A risk trigger is an occurrence or particular event that is likely to result in an identified risk materialising. When a trigger event occurs, it enables the risk manager or project manager and other project team members to initiate a response to an identified risk with a view to mitigating the potential impact.

Relationship to other risks (interdependencies)
In any construction project it is extremely rare that any activity is totally independent of the activities which occur either concurrently or consequentially, and this will almost always be the case for risk.

Whilst it may be acceptable to ignore interdependencies when undertaking quantitative analysis (i.e. when allowances for risks are quantified – see later chapter) it is essential to the successful implementations of RM that the relationship between risks is appreciated and understood. Indeed the situation may often occur where one risk may only arise if another risk or activity has caused it to do so.

Furthermore, the response to one risk may give rise to a resultant or consequential secondary risk, hence the need for the constant iterative nature of the RM process.

Potential impact
Impact is usually measured in terms of time, cost and quality. At this early qualitative assessment stage the information is unlikely to be available to accurately predict the potential impact of the risk event on these factors.

However, even at this initial stage, the potential impact of the risk events should be classified into HIGH, MEDIUM and LOW categories to enable the potentially high-impact risks to be given more fundamental consideration than those with negligible impact.

As previously stated all impacts will generally have an effect on time, cost and quality. But as any impact on time or quality is more usually associated with a form of financial loss, it is prudent to state that all risk impacts can be presumed to have a cost effect.

Likelihood of occurrence (probability)

As with potential impact it is unlikely that at this initial assessment stage a detailed analysis of the likelihood of occurrence is a practical exercise. However, a subjective assessment which categorises the likelihood (probability) of occurrence based on intuition and experience should be undertaken. This may be given in different formats as follows:

- probability 0 – 100%
- low/medium/high.

Response/mitigation

This is the action required to reduce, eradicate or avoid the potential impact of risks on a project. For each risk that is identified the risk or project manager must consider alternative methods to mitigate the risk of using the most appropriate process. Risk response/mitigation is discussed in more detail in Chapter 5 but can be briefly summarised as follows:

- *avoidance* this process is concerned with removing the cause of the risk by investigating what factors are likely to cause it to materialise and attempting to eradicate them. This is synonymous with rejecting a risk; the most simple form is not to tender for a project.
- *transfer* this is basically the process of transferring the risk to those best-placed or more capable of managing them. This may mean transferring the risk to a specialist sub-contractor who understands the subject better and is more capable of identifying and managing the risk. However, there may be

cost implications with this process as sub-contractors may increase their costs to cover these risks.

- *reduction*   this process involves the reduction of risk by sharing the risk with another party. This may be in the form of taking out an insurance policy or capital investment in equipment to attempt to prevent a risk event occurring, e.g sprinkler system in a building to lessen effect of fire.

  The purpose of this process is to take an action which will result in the risk being reduced to an acceptable level. This process generally involves the expenditure of monies, e.g. an insurance premium, employing an independent quality checking company to obtain a supplementary check on a company's projects and education and training of staff to alert them to the potential of risk.

- *risk responsibility*   allied to the classification of risk it should be possible to identify that person or party who is best placed to influence and therefore deal with such risk. In essence any successful risk management regime will allocate responsibility for each risk to the most appropriate person or party who has the required level of authority to act and respond to that risk in order to mitigate its potential impact. That person will also remain responsible for reporting on necessary action taken, results achieved and consequential risks that may have arisen directly to the Project Manager or Risk Manager.

Once this initial qualitative assessment has been completed it may be possible to prioritise risk using categories of potential risk exposure, e.g.

unacceptable
critical

significant
insignificant
acceptable.

To quickly identify risk exposure a simple matrix can be used as shown in Figure 16.

Impact

High	significant	critical	unacceptable
Medium	insignificant	significant	critical
Low	acceptable	insignificant	significant

Likelihood

Low		Medium		High

Figure 16. *Risk exposure matrix*

The above exercise would allow the project team to focus their attention upon those risks that are likely to cause the greatest concern. It may become clear, even at this early stage, that the overall level or risk is unacceptable and therefore needs to be significantly redirected or even aborted.

Once the qualitative risk assessment is complete all risks should be transferred to the risk register (see Figure 17) which is then used to actively review risks and updated to reflect action required or taken and to include any new or resultant risks which may arise. Risk identification is of paramount importance and should be approached in a formal manner.

The problems associated with relying on the intuition of experienced project members is that they have a tendency to concentrate on their own specialisations and experience and lateral thinking is not promoted.

Experienced project members have proved unreliable because these members are generally too busy or if they are totally involved with a particular project they tend to play down any risks as they feel they have seen it all before.

There is also a reluctance to discuss risks with other members outside of their specialised fields. People tend to believe that

Ref.	Brief description	Trigger Event	Interdependencies (Ref to)	Pb	Effect	Impact ARA	MLRA	Response/mitigation Status Current	Previous	Owner	Status	Comments/ actions

T = Third party risks
S = Site risks
Cl = Client risks
D = Design team risks
Co = Contractors risks
O = Other risks

*Ref to is the
reference to the
dependent risk(s)

T = Time
C = Cost
Q = Quality

A = Assessed
M = Managed out
D = Designed out
S = Shared/transferred
I = Ignored

I =- Initial/intuitive
C = Considered
F = Final

Figure 17. *Risk register*

risks belong to them personally and do not realise that the risks can affect other members.

Unfortunately, when firms are bidding for work in competition, as is the case in most tendered construction contracts, there may be a tendency to understate the risk or its effects or even an unwillingness to raise the possibility or risk so as to gain an advantage over other tenderers.

It is of much greater benefit to be able to respond to risks rather than having to deal with their consequences when they materialise unexpectedly.

5
Response and mitigation

Introduction

Risk response and mitigation is the action that is required to reduce or eliminate the potential impact of risk. There are two types of response to risk – one is an immediate change or alteration to the project which usually results in the elimination of the risk; second is a contingency plan that will only be implemented if an identified risk should materialise.

In order to mitigate the potential impact of any risk the project manager or his designated risk manager must consider alternative courses of action and evaluate the consequences should that action be taken. As an integral part of the RM process, the main aim of any response and mitigation strategy is to initiate and implement appropriate action to prevent risks from occurring or, at minimum, limit the potential damage they may cause.

Furthermore, through the use of adequate and appropriate contingency plans, if the occurrence of a risk is unavoidable, its impact should be limited to the contingency levels contained within the overall project allowances. This should ensure that the overall project objectives of time, cost and quality are not jeopardised.

These response and mitigation strategies may be formulated to respond to individual or groups of risk as deemed appropriate by the project/risk manager. However, strategies must be agreed, recorded and documented with all responsibilities clearly stated. In particular, it is clear that high probability/high impact risks will demand rigorous and thorough discussion and examination to ensure the formulation of an appropriate and credible response and mitigation strategy.

The options for responding to risk are avoidance, reduction, transfer, sharing and retention - each should be assessed as one or more will aply in every circumstance. In order to identify which route(s) should be adopted a number of questions must first be asked:

- is the risk controllable or uncontrollable
- who is best placed to influence/deal with the source and outcome of the risk
- what secondary or resultant risks arise as a result of the action taken
- is the cost of mitigating the risk acceptable when compared to the potential impact of the risk itself?

Whilst detailed quantitive assessment is explained in more detail later in this chapter it is worth appreciating that:

Potential   = *Likelihood*   *x*   *consequence or effect*
Impact   *or probability*   *of risk should it materialise*

Risk avoidance
Risk avoidance may include a review of the overall project objectives leading to a reappraisal of the project as a whole. Risk avoidance is often perceived as the ultimate mitigation strategy in that it implies that the project may be aborted.

In simple terms, this method of mitigation involves the removal of the cause of the risk and therefore the risk itself. Ideally any approach involving avoidance is best implemented by the consideration and adoption of an alternative course of action. Other examples of risk avoidance include the use of exemption clauses in contracts, either to avoid certain risks or to avoid certain consequences folwing from the risks. Risk avoidance is most likely to take place where the level of risk is at a level where the project is potentially unviable.

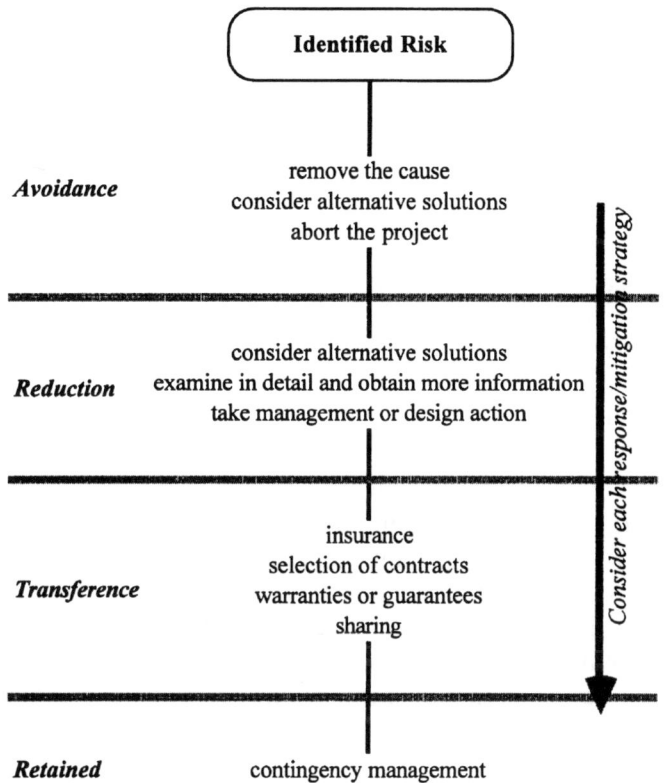

Figure 18. *Considering risk responses*

Risk reduction

This method adopts an approach whereby potential exposure to risks and their impact is alleviated. Often this is achieved by the managing or designing out of potential risk. Methods of risks reduction may require some initial investment which should then reduce the likelihood of the risk occurring. Risk reduction occurs where the level of risk is unacceptable and alternative action is available. Typical action to reduce risk could be:

- detailed site investigation where adverse ground conditions are known to exist but the full extent is not known; a detailed ground investigation will improve the information upon which the estimate has been prepared

- alternative procurement route – by utilising an alternative contract strategy risks will be allocated between project participants in a different way
- changes in design to accomodate the findings of the risk identification process.

Risk reduction exercises will always be worthwhile because they can lead to greater knowledge about the project and this reduces not only the potential impact of risks but also the level of uncertainty – itself a major source of risk. Risk reduction invariably leads to greater confidence regarding the project's outcome. However, risk reduction will result in an increase in the base cost (i.e.the estimate of all certain items) but should offer a significantly greater reduction in the level of contingency required. It goes without saying that risk reduction should only be adopted where the resultant increase in costs is less than the potential loss that could be caused by the risk being mitigated.

It is worth noting that some texts view insurance as risk reduction whilst others consider it as a transference of risk. Neither is incorrect, the most important issue is that the risk is dealt with through an appropriate response/mitigation.

Risk transfer

This method involves the transfer of risk to other project participants. Commonly, risks are transferred through the placement of contracts, the appointment of specialist sub-contractors or suppliers or by taking out an insurance policy. As with risk reduction, whenever a risk is transferred there is usually a premium to be paid. Again, this premium should be less than the potential impact of the risk.

Tranference of risk should comprise the passing of risks to those better placed or more capable to maintain control or influence the outcome of the risk. Transference should never be viewed as a negative risk response. Its intention is not to pass the buck by making someone else responsible. Further, it should not be used in a penal or punitive manner as a protective mechanism for other project participants. For risks to be managed properly an incentive may be required.

When transferring risk it is important to differentiate between the transference of the risk itself and the allocation of risk responsibility. Where a risk is transferred the intention should be to transfer the whole of the risk including its potential impact. Where the responsibility for the risk is allocated to a project participant, time, cost and quality repercussions remain and this may still adversely affect the project's outcome.

Where a portion of the risk is transferred whilst some risk is retained this is known as risk-sharing. This approach may be adopted where the risk exposure is beyond the control of one party. In such instances it is imperative that each party appreciates the value of the portion of risk for which it is responsible.

A typical example of the transfer of risk would be in the selection of procurement routes as indicated below.

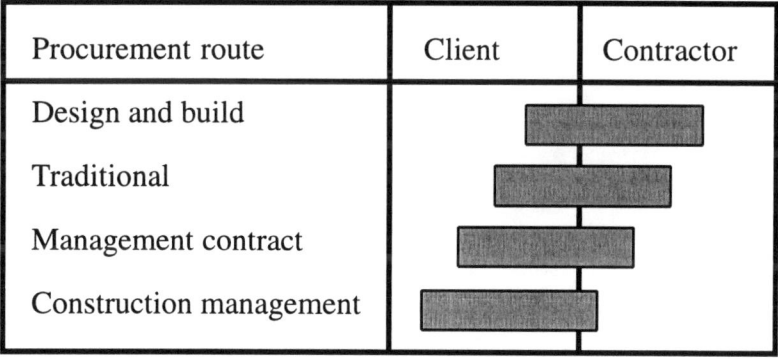

Procurement route	Client	Contractor
Design and build		
Traditional		
Management contract		
Construction management		

Figure 19. *Allocation of risk in procurement routes*

Residual or retained risk
Once all the avenues for response and mitigation have been explored a number of risks will remain. This does not imply that these risks can be ignored, indeed it is these risks which will in most instances undergo detailed quantitative analysis in order to assess and calculate the overall contingency levels required. The aim of the previous responses is to reduce project

uncertainty and in so doing increase the base estimate to reflect the more certain nature of the project. However, it does not imply that these retained risks can simply be ignored. Indeed, they should be subject to effective monitoring, control and management to ensure they are contained within the contingency allowances set.

It should be noted that this contingency should be made up of residual risks which are assessed to be of a low likelihood and low potential impact. High probability and high impact risks should undergo further rigorous examination so that an alternative response can be found.

6
Detailed quantitative assessment of risk (risk analysis)

Introduction

The detailed quantitative assessment of risk is the one most people identify as risk analysis. In undertaking a quantitative assessment you will attempt to apply meaningful and objective probabilities to risks and subsequently consider and then quantify the potential impact of such risks in terms of time, cost and quality.

Quite clearly whilst impact on quality can be established it cannot be quantified as such. With this in mind the remainder of this section will consider quantitative assessment from the viewpoint of converting identified risks into realistic allowances for cost and time that should then be added to a risk free base.

In considering the potential impact of risks it is essential to appreciate that when risks do materialise, they do not all materialise at once or to their full potential. In particular, the fact that the risk has been identified usually means that some form of mitigation is in place. The potential impact of risks are, therefore, often measured on a swings and roundabouts basis in that some risks will materialise while others will not. Clearly, if the worst case was considered in every situation most projects would not proceed!

When preparing any estimate it can generally be split into

two distinct elements, namely:

- a base estimate   an estimate of all those items which are known or a degree of certainty exists
- a contingency   an allowance for all the uncertain elements of a project.

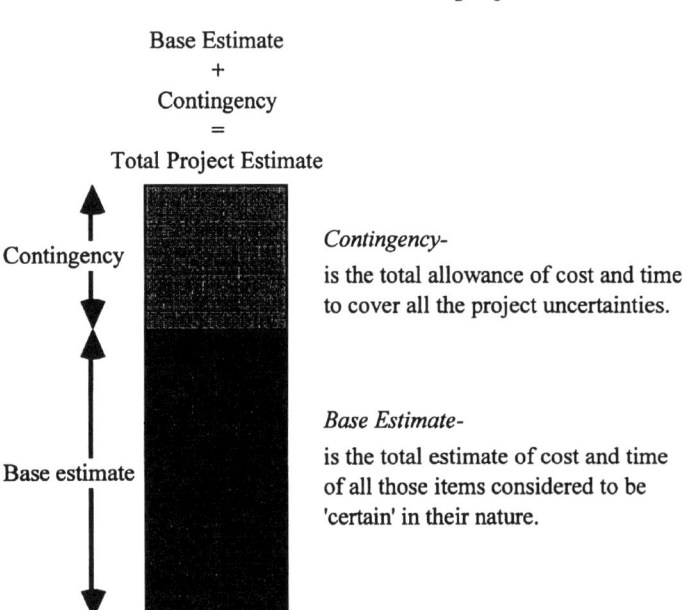

Base Estimate
+
Contingency
=
Total Project Estimate

Contingency

Base estimate

Contingency-
is the total allowance of cost and time to cover all the project uncertainties.

Base Estimate-
is the total estimate of cost and time of all those items considered to be 'certain' in their nature.

Figure 20. *Build up of estimates*

Historically, contingencies have often been calculated on a rule of thumb basis, e.g. 5% for new build and say, 10% for refurbishment projects. By adopting a risk management approach, contingencies are set up to reflect realistically the risks inherent in the project. Used correctly, contingency allowances facilitate an improvement in the overall accuracy of estimates and ensure that expenditure against risks is controlled.

If the base estimate plus a contingency reflects the expected outcome of a project it would be useful to have some degree of confidence in the estimate being reported and upon which decisions are made. In most instances the expected outcome is considered as the most likely. Analysing this in detail tends to indicate that this estimate will have a 50/50 success rate, i.e. a 50% chance that the outcome will be below the estimate and 50% chance that it will be above. To some clients this may be acceptable, but to many it is not. Perhaps more importantly is the difference between this 50% confidence and the estimate at say a 95% confidence limit.

If the difference between the two is too great then the degree of risk may be too high and an alternative course of action or a detailed review of the projects status may be appropriate. Calculating these confidence limits will be dealt with in more detail later in this chapter.

As any project progresses, estimates contain more certainties then uncertainties and therefore the contingency allowance can be reduced. If, as stated earlier, risk management aims to ensure that projects are delivered at the lowest cost, compatible with the specified quality and in the desired time, then the expected outcome should be consistent while the base estimate increases and the level of tolerance to, say a 95% level decreases.

In Figure 21 it can be clearly seen that as the project develops so its outcome becomes more certain.

The aim of this quantitative analysis is therefore to set a contingency at a level which realistically reflects the risk at that time. The size of the contingency should be based upon the results of the formal risk analysis, if it is too high the project may not proceed, too low and the project could be compromised at some later date. In addition to calculating this contingency it is also necessary to arrive at a tolerance which will reflect the difference between the most likely outcome (i.e base + contingency) and a higher level of confidence e.g. 95%.

The remainder of this chapter deals with the various methods of quantifying risk. It is not intended to advocate one method over another but to simply give a brief introduction into the various methods used. In addition there are examples of some of the simpler methods that may be adopted for smaller less complicated projects where the degree of interdependency may be small. The methods of quantitative assessment discussed are:

- simple assessment

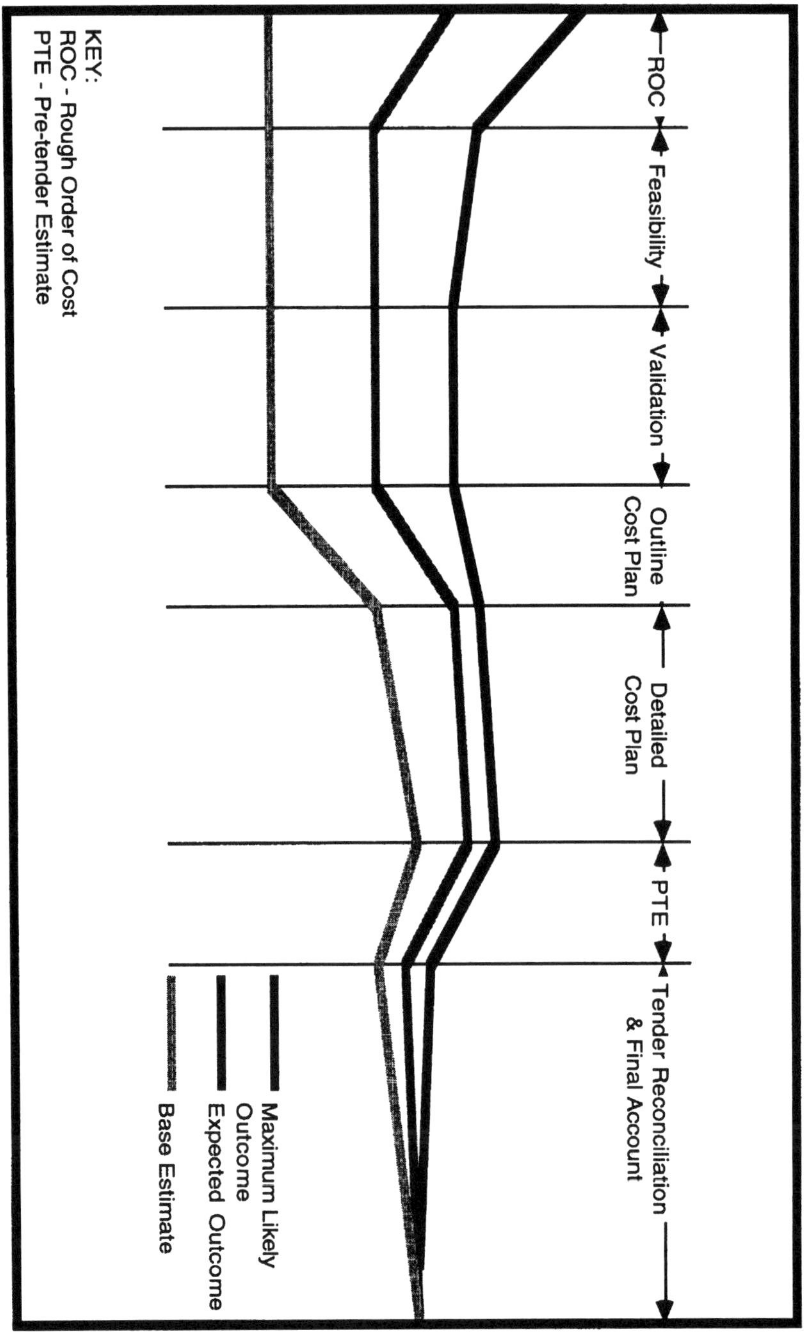

Figure 21 *Cost convergence*

- probabilistic analysis
- MERA (multiple estimating using risk analysis)
- sensitivity analysis
- decision trees
- stochastic dominance
- Monte Carlo Simulation and Latin Hyper Cube Sampling.

However, prior to looking at the various techniques in detail it is necessary to have a basic knowledge of statistics.

A brief introduction to statistics

It is important not to be overly concerned with the terminologies involved but an attempt to understand some of the basic principles of statistics will be most useful.

Descriptive statistics	this is described as the analysis of past events in order to identify means, medians, variances and deviations. From an estimating point of view this would be seen as the cost analysis stage. Descriptive statistics provides much of the background information which enables inferential statistics to take place.
Inferential statistics	can be described as a method of making decisions based upon incomplete information. This method of statistics is, therefore, the prime area around which risk analysis is based. Inference is usually based upon a process of random sampling of known factors or information.
Mid-range	the mid range is that number which lies halfway between the highest and lowest observations, e.g.

take the numbers 2,4,6,8,10

$$\text{mid-range} = \frac{2+10}{2} = 6$$

Mode   the mode is defined as that number that occurs most frequently. If each number occurs the same number of times then no mode exists, e.g.

2,4,6,8,10   – no mode exists
2,4,6,6,8,20   – then the mode is 6
2,2,4,6,6,8,9   – this is bi-modal, 2 & 6

Median   if there are a number of observations and these are arranged in order then the observation that lies in the middle is known as the median, e.g.

8,10,6,4,2 in order is 2,4,6,8,10 and the median is 6

if there are an even number of observations then the median is the number between the two middle observations, e.g.

2,4,6,8,10,12 the median is 7, i.e. between 6 and 8.

Mean   Mean is often known as the average and is calculated by summing the results and dividing by the total number of observations, e.g.

$$\frac{2+4+6+8+10}{5} = 6$$

in some instances the upper and lower limits are excluded from this calculation if they are considered too extreme.

Distributions   the distribution of a number of observations can be shown in a number of formats. Using the example of rolling two dice one hundred times the results are as follows:

Observation	Frequency (no.of observations)	Cumulative frequency
2	1	1
3	6	7
4	4	11
5	14	25
6	15	40
7	17	57
8	10	67
9	14	81
10	10	91
11	8	99
12	1	100

This information can be provided in a number of formats such as histograms, cumulative frequency polygons or distribution curves.

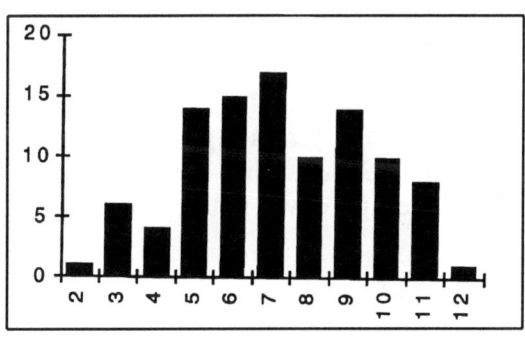

Figure 22. *Histogram*

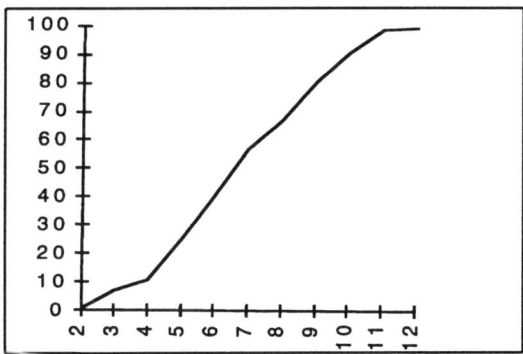

Figure 23. *Cumulative frequency polygon*

Figure 24. *Distribution curve (indicating normal distribution)*

Confidence limits

If the exercise of rolling two dice was carried out on a large scale, say 100,000 times then the distribution would be more symmetrical in shape. This is known as normal distribution and is indicated in Figure 24 above. the point at which there is confidence that a value or cost will not be exceeded. This is expressed in percentage terms e.g. a confidence limit of 90% implies in 9 cases out of 10 the figure will not be exceeded.

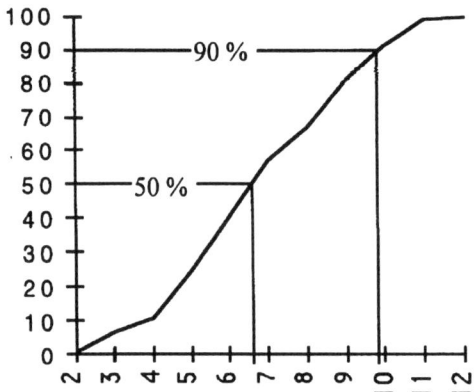

Figure 25. *Confidence limits*

Probability

probability is the likelihood of something happening. If a coin is tossed it will either land heads or tails. The likelihood or probability of it landing heads is 50% or 0.5, the same is true for tails. Probabilities can be either independent (i.e. do not rely on the outcome of a previous event, e.g. tossing a coin – it is irrelevant if the last time the coin was tossed it landed heads or tails) or dependent (i.e. where the outcome may be reliant on other events e.g. if there are three pairs of blue and three pairs of black socks in a drawer and a blue pair has been removed then the probability of picking a black pair next time is increased due to the previous event).

Where there are two or more events then the probabilities of each are multiplied together to give the overall probability of getting the combination. Consider this - using two coins, what is the probability of getting two heads:

Probabilty (Pb.) of 1 head   = 0.50
Pb. of another head   = 0.50
Pb. of two heads, 0.50 x 0.50 = 0.25

This can be expanded upon:

Coin A	Pb	Coin B	Pb	Combined
Heads	0.5	Heads	0.5	0.25
Heads	0.5	Tails	0.5	0.25
Tails	0.5	Heads	0.5	0.25
Tails	0.5	Tails	0.5	0.25
			Total	1.00

The sum of the probabilities *will always be one*.

Methods of risk assessment

This chapter will now look in detail at some of the methods of quantifying risk. The usefulness of having a brief knowledge of statistics will soon become apparent. To those who wish to venture into some of the mathematical-based methods of assessment it may be appropriate to purchase an introductory text into statistics which will explain some of the principles in more detail.

Simple assessment

This is a relatively simple arithmetical, as opposed to a more complex statistical, technique which considers significant risks separately and examines the possible combined effect. An estimate is prepared calculating the expected impact of each significant risk which are then added together to give an approximation of the overall level of risk allowance required. This technique is generally only suitable to those projects which are both small in value and relatively simple in their nature.

For example, a small new-build scheme (say £120,000) has a risk of requiring a drainage diversion at a cost of approximately £10,000. This should be added in full to the base estimate to provide expected outcome because there are likely to be few risks of this magnitude on such a small scheme. It is therefore a simple financial assessment of risk.

This simple assessment may be enhanced further by following the procedure of measuring the potential impact of the risk by multiplying its probability or likelihood of occurrence by its full consequence if it did occur. This can only be used if a reasonable number of risks exist as it assumes that some risks will materialise and others will not, i.e. a swings and rounabouts approach. Clearly the danger with this is that if there is a high impact risk which does materialise it could wipe out any contingency provisions immediately.

Probabilistic analysis

The method of probabilistic analysis relies upon applying meaningful, although somewhat subjective, probabilities to estimates. However, detailed analysis of historic costs and their spread or range should enable the approach to be more scientific and less arbitrary. This is a statistical technique which enables the calculation of cost or time consequences for either individual risks (which can then be combined) or for the project as a whole. To undertake this exercise, risks are expressed as a range of probabilities with their associated values. Normally each important risk can only realistically be quantified in terms of optimistic, most likely or pessimistic estimate to which probabilities are then applied. It should be noted that the number of intervals is not fixed at three but may be as many as is deemed appropriate for each particular project. The sum of these probabilities should equate to 1 or 100%.

The estimate is then multiplied by its probability and the sum of the results is expressed as the Expected Value (EV) or is sometimes termed Average Risk Estimate (ARE).

An early cost estimate for an as-yet undesigned office building.

Estimated Costs	Probability	Result
£500/m^2	30% (0.3)	150
£750/m^2	60% (0.6)	450
£1000/m^2	10% (0.1)	<u>100</u>
	Expected Value	£700/m^2

The maximum likely risk estimate is the pessimistic estimate, i.e. in this instance £1,000/m^2.

As stated previously the method highlighted above results in an Expected Value and is sometimes known by this name. Discrete Probabilistic Analysis is an extension to this and can be used to combine risks by the process of independent addition. For example, two different elements of a project may have differing probabilities of differing outcomes e.g. floor and ceiling finishes:

Floor finishes		Ceiling finishes	
Cost	Probability	Cost	Probability
10,000	0.1	10,000	0.3
20,000	0.5	15,000	0.6
30,000	0.4	20,000	0.1

Rather than summing the estimated values these can be combined as follows:

Floor finishes		Ceiling finishes		Cost	Probability
Cost	Pb	Cost	Pb	(Floor + Ceiling)	(Floor X Ceiling)
10k	.1	10k	.3	20k	.03
		15k	.6	25k	.06
		20k	.1	30k	.01
20k	.5	10k	.3	30k	.15
		15k	.6	35k	.30
		20k	.1	40k	.05
30k	.4	10k	.3	40k	.12
		15k	.6	45k	.24
		20k	.1	50k	.04

It is important to remember that when using this method of combining expected values, the events or estimates are added together whilst the probabilities are multiplied. Therefore a table can be produced by adding together the probabilities for the same costs, e.g. 30k appears twice with a probability of .01 and .15, the total probability of 30k is therefore .16.

Cost (Floor and Ceiling)	Computation	Total probability
20k	.1 x .3	.03
25k	.1 x .6	.06
30k	(.1 x .1) + (.5 x .3)	.16

Cost	Computation	Total
(Floor and Ceiling)		probability 35k
	.5 x .6	.30
40k	(.5 x .1) + (.4 x .3)	.17
45k	.4 x .6	.24
50k	.4 x .1	.04

Combined Expected Value is therefore:

Cost	Probability	Result
20,000	.03	600
25,000	.06	1,500
30,000	.16	4,800
35,000	.30	10,500
40,000	.17	6,800
45,000	.24	10,800
50,000	.04	2,000

Combined Expected Value   £37,000

This technique of combining values by independent addition does not, however, take account of dependencies, i.e. where the outcome of one event impacts on the outcome of a subsequent event. In order to do this we must attempt to understand the causal effect where dependencies exist:

Screed	Pb	Floor finish	Pb
3,000	.2	5,000	.4
		6,000	.6

_____ Line indicates dependencies to previous choice.

4,000	.3	3,000	.3
		4,000	.3
		5,000	.3

_____

Screed	Pb	Floor finish	Pb
5,000	.5	3,000	.3
		4,000	.4
		5,000	.3

Here the type of vinyl floor covering used may be dependent upon the screed used. Utilising the same technique as before:

Screed	Pb	Floor finish	Pb	Combined	Pb
3,000	.2	5,000	.4	8,000	.08
		6,000	.6	9,000	.12
4,000	.3	3,000	.3	6,000	.09
		4,000	.3	8,000	.09
		5,000	.3	9,000	.09
		6,000	.1	10,000	.03
5,000	.5	3,000	.3	8,000	.15
		4,000	.4	9,000	.20
		5,000	.3	10,000	.15

Summing as before:

Cost	Probability	Result
7,000	0.09	630
8,000	0.32	2560
9,000	0.41	3690
10,000	0.18	1800
	Estimate	£8680

Whilst this may be relatively straight forward when looking at two elements only, it is a difficult task to undertake manually when many dependents exist. Monte Carlo Simulation, which is discussed later, overcomes this by simulating elements and combining them statistically.

Advantages of probabilistic analysis

- simple to apply and understand
- gives some indication of the range of possible outcomes and their probabilities.

Disadvantages of probabilistic analysis

- an amount of subjective estimation is necessary.

Figure 26 overleaf indicates a Typical Probabilistic Estimate

Multiple estimating using risk analysis (MERA)
Multiple estimating using risk analysis (MERA) is a technique devised by the PSA which attempts to provide a range of estimates. These are presented as a risk-free Base Estimate, an Average Risk Estimate (ARE) and a Maximum Likely Risk Estimate (MLRE).

The ARE is the risk free Base Estimate plus an average risk allowance and the MLRE is the ARE plus the maximum risk allowance. For example:

Risk-free Base Estimate	10,000,000
Calculated Average Risk Allowance	1,000,000
Average Risk Estimate (ARE)	£11,000,000
Calculated Maximum Risk Allowance	3,000,000
Maximum Likely Risk Estimate (MLRE)	£14,000,000

As with probabilistsic analysis, MERA attempts to find a level that is the most likely outcome, i.e. that estimate which has a 50% chance of being successful. This is the Average Risk Estimate (ARE) and could be equated to the Base Cost plus contingenecy explained at the beginning of this chapter. The Maximum Likely Risk Estimate (MLRE) is that estimate which

ELEMENTAL COST PLAN

Notional Cost Plan Using Probabilistic Analysis

Elements	Optimistic Cost	Pb.	Realistic Cost	Pb.	Worst Cost	Pb.	Expected Value
1 SUBSTRUCTURE	25,000	10%	27,000	50%	30,000	40%	28,000
2 SUPERSTRUCTURE							
A Frame	15,000	20%	18,000	60%	21,000	20%	18,000
B Upper Floors	1,500	20%	2,000	70%	2,500	10%	1,950
C Roof	12,000	30%	15,000	50%	17,500	20%	14,600
D Stairs	500	15%	750	45%	850	40%	753
E External Walls	17,500	20%	20,000	65%	22,000	15%	19,800
F Windows and External Doors	6,500	40%	7,000	50%	9,000	10%	7,000
G Internal Walls and Partitions	3,000	30%	4,000	40%	5,000	30%	4,000
H Internal Doors	4,500	20%	6,000	60%	7,000	20%	5,900
Total Element Group Cost							
3 INTERNAL FINISHES							
A Wall Finishes	7,500	40%	8,500	55%	10,000	5%	8,175
B Floor Finishes	3,000	10%	4,000	70%	5,000	20%	4,100
C Ceiling Finishes	2,500	20%	3,000	65%	4,000	15%	3,050
Total Element Group Cost							
4 FITTINGS AND FURNISHINGS	7,000	25%	8,000	50%	10,000	25%	8,250
Total Element Group Cost							
5 SERVICES							
A Sanitary Appliances	1,000	30%	2,000	50%	2,500	20%	1,800
C Disposal Installations	750	30%	1,000	40%	1,200	30%	985
E Mechanical Installations	25,000	20%	30,000	50%	35,000	30%	30,500
H Electrical Installations	30,000	10%	35,000	60%	40,000	30%	36,000
N BWIC with Services	2,000	20%	3,000	70%	5,000	10%	3,000
Total Element Group Cost							
Sub-total excluding external works, preliminaries and contingencies	164,250		194,250		227,550		195,863
6 EXTERNAL WORKS							
A Site Works	30,000	30%	40,000	50%	50,000	20%	39,000
B Drainage	15,000	20%	20,000	60%	25,000	20%	20,000
C External Services	3,000	30%	4,000	40%	5,000	30%	4,000
Total Element Group Cost							
7A PRELIMINARIES 10.00%	21,225		25,825		30,755		25,886
8A CONTINGENCIES 5.00%	11,674		14,204		16,915		14,237
							298,986

Figure 26. *Typical probabilistic estimate*

has a nine in ten or 90% chance of not beeing exceeded and is often defined as the Base Cost plus Contingency plus Tolerance. This method of costing construction projects is the one adopted by the Treasury and a number of other Government bodies.

In order to calculate the average risk allowance and maximum risk allowance we must first identify those risks which are of a fixed or variable nature:

Fixed risk — a risk which will either be incurred as a whole or not at all, e.g. a new gas main – it will either be required or not required

Variable risk — a risk relating to a circumstance which has a varying probability of occurrence with a variable outcome, e.g. additional parking – additional parking may be required but the extent is unknown.

Those risks now identified as fixed should be multiplied by their probability of occurrence to calculate the average risk allowance; the maximum risk allowance is the full value of that risk. The probability of occurrence is assessed using, for example, the categories outlined below.

Likelihood	Mathematical factor
Nil chance or probability	0
Unlikely or improbable	0.05– 0.45
	(5%–45%)
As likely as not	0.45–0.55
	(45%–55%)
Likely or probable	0.55–0.95
	(55%–95%)
Certain	1
	(100%)

For example: for the requirement of a new gas main:

Probability of occurrence	Average of Risk Allowance	Maximum Risk Allowance
70% (or 0.70)	70,000 (100,000 x 0.70)	100,000 (i.e. full cost of risk)

The average risk and maximum risk allowances are now also calculated for variable risk items. The average risk allowance for a variable risk is that sum estimated as having a 50% probability of covering the cost of the risk. The maximum allowance is that sum which has a probability of 90% of covering the cost of the risk (although this is not sacrosanct and may range from 80% to 100%). This method of calculation may be seen as rather subjective. However, with thought and considered assessment it should be possible to achieve a meaningful estimate.

Unlike Fixed Risks the probability of a Variable Risk occurring does not directly impact on the Average and Maximum Risk Allowances inasmuch as the allowances are calculated around confidence or prediction limits and not derived from the probability factors.

Fixed Risk; Estimate of risk = 100,000 and probability of occurrence = 0.6, therefore average allowance is 60,000 (100k x .6) and maximum allowance = 100,000 (i.e. full cost of risk).

Variable Risk – may also have a probability of occurrence of 0.6 but the average and maximum risk allowances are calculated thus the graph in Figure 27 indicates assumed distribution of costs.

The maximum allowance is that point where 90% of the area under the curve lies to the left, i.e. shaded area or using a cumulative graph (known as a cumulative frequency polygon).

The points measured above are the 90 percentile along the Y axis measured across to where it cuts the cumulative frequency curve. If the maximum allowance multiplied by a probability of 0.9 was used this would equate to measuring along the X axis and would, therefore, give an answer which is 9/10 of the maximum as opposed to 9/10 of the cumulative frequency.

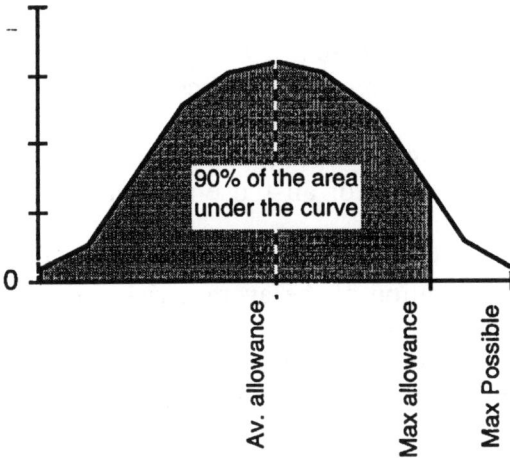

Figure 27. *Risk allowances for a variable risk*

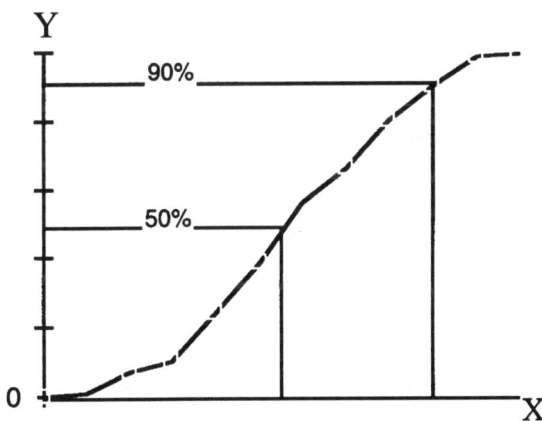

Figure 28. *Frequency polygon for variable risk*

Risk	Average Risk Allowance	Maximum Risk Allowance
Possible requirement for additional parking spaces	£50,000	£200,000

In this example it could be said that there is a 50% probability that 50 additional spaces at £1,000 per space will be required –

hence average risk allowance = 50 x £1,000 = £50,000. The maximum risk allowance is that there is a 90% probability that no more than 200 spaces will be required – hence 200 x £1,000 = £200,000.

This process of calculating an average and maximum risk allowance should be completed for all those risks identified during the risk identification stage.

Having calculated the risk allowance it should be noted that the key factor in the preparation of the average risk estimate and the maximum likely risk estimate using the MERA technique is that risks are not included at their full potential on the basis that some will occur whilst others will not (i.e. the likelihood of them all occurring is extremely remote). Therefore, a statistical technique has been devised for combining all maximum likely and average costs to produce a meaningful estimate.

This is achieved by summing the squares of the spread of each risk and square rooting the total, the resultant being added to the average risk estimate.

Taking this one step at a time:

Summing the squares of the spread (the spread is the difference between the maximum and average risk allowance).

Risk	Average Allowance	Maximum Allowance	Spread (Max less average)
Gas Main	70,000	100,000	30,000
Parking	50,000	200,000	150,000

Now the spreads are squared

$$30,000 \text{ x } 30,000 \quad = 0.09 \text{ x } 10^{10}$$
$$150,000 \text{ x } 150,000 \quad = 2.25 \text{ x } 10^{10}$$
$$\text{And sum them} \quad = 2.34 \text{ x } 10^{10}$$

Next step is to square root the total and add it to the average risk estimate.

Assuming base cost of £1,000,000. The average risk allowance is the sum of all the individual average risk allowances which in this instance is:

70,000 + 50,000	<u>120,000</u>
Average Risk Estimate	1,120,000
Now add the square root of the sum of the	
spreads squared _2.34 x 10^{10} =	<u>152,970</u>
Maximum Likely Risk Estimate	<u>£1,272,970</u>

Therefore the final schedule would look something like:

Risk	Probability of occurrence	Average risk allowance	Maximum risk allowance	Spread (ie max less av.)	Spread^2
New gas					
main	70 %	70,000	100,000	30,000	0.09 x 10^{10}
Parking		<u>50,000</u>	<u>200,000</u>	150,000	2.25 x 10^{10}
		120,000	Sum of the squares	= 2.34 x 10^{10}	
			Square root	= £ 152,790	
			Base cost	£1,000,000	

Average Risk Estimate	= £1,000,000 + £120,000 = £ 1,120,000
Maximum Likely Risk	
Estimate	= £1,120,000 + £152,970 = £1,272,970

Clearly this exercise is preferably carried out using computer technology (e.g. spreadsheets) although it can be executed manually.

Advantages of MERA

• relatively simple to understand and apply
• produces meaningful estimates.

Disadvantages of MERA

- due to the statistical techniques involved, the Maximum Likely Risk estimate could be exceeded.

Sensitivity analysis

Sensitivity analysis is a practical method of showing the effects of risk or uncertainty on the project by varying the values of key factors and measuring the outcome. This method does not use subjective probability estimates but aims to provide estimate-based information upon which decisions may be made. As such it does not furnish mathematical results which are themselves used in the preparation of estimates, but highlights key factors which could have a significant impact on the overall project should they be varied. In doing so, sensitivity factors can be calculated.

A sensitivity factor is calculated by dividing the percentage change in the outcome by the percentage change in the base element, e.g.

Assumed estimated cost	250,000	
Inflation allowance 5%		<u>12,500</u>
		<u>£262,500</u>

If there was a change in the inflation rate from 5% to 6% (i.e. a 20% increase in the rate)

Estimated cost	250,000
Inflation allowance of 6%	<u>15,000</u>
	<u>£265,000</u>

Therefore the sensitivity factor is calculated as follows:

265,000 – 262,500 = an increase of 2,500
2,500/265,000 = an overall percentage increase of 0.94%
0.94/20% = a sensitivity factor of 0.05

Sensitivity factors themselves furnish little information unless they are measured against other elements. However, this is a somewhat pointless exercise when comparing capital costs because the effect of the change is easily calculated on the overall summary. This method is normally used when calculating investment values or net present values where changes in items such as yield, interest rates, and inflation have significant effect on the overall outcome.

One use in project estimates could be for inflationary allowances where 'what if' scenarios can be run using a simple spreadsheet, e.g.

Using the example above

if the allowance is	5%	outcome =	262,500
	6%	=	265,000
	7%	=	267,500
	8%	=	270,000 etc.

Much of this is common sense but the highlighting of those items which are volatile and have a higher sensitivity factor enables a concentration of effort on those elements which have a greater overall impact on the project.

Advantages of sensitivity analysis

- indicates the effect on the outcome of the project when varying the value of the elements which make up the project
- simple in principle
- enables the identification of the most sensitive elements.

Disadvantages of sensitivity analysis

- gives no indication of the likelihood of a variation occurring.

Decision trees

Decision trees are a pictorial method of showing a sequence of inter-related decisions highlighting possible courses of action and future possible outcomes. Where probabilities and values of potential outcomes are known, or can be estimated, they are used as a method of quantification to provide a more informed basis for notional decision making. The aim of the decision tree is to produce an Expected Value (EV) for each option in

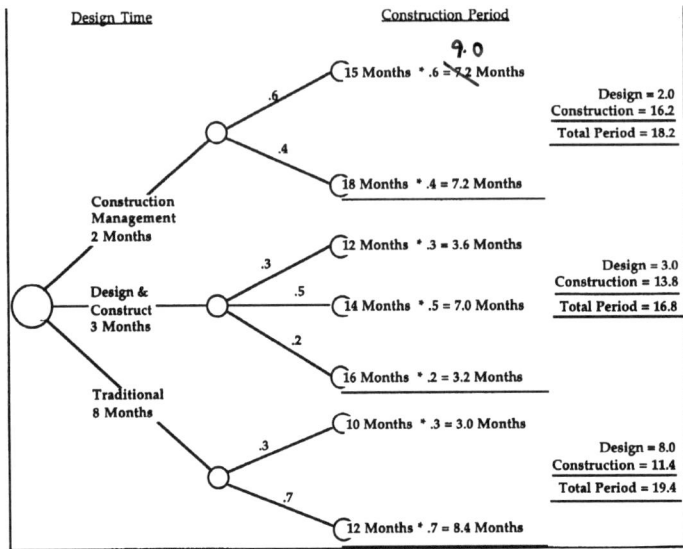

Figure 29. *Typical decision tree*

the decision making process.

The figure above indicates how alternative procurement routes can be examined and compared in order to identify which option is likely to be completed in the shortest timescale.

Bayesian theory or Bayes' rule is a refinement of the decision tree technique by revising estimated probabilities in the light of additional information thereby enabling new information to be incorporated into the analysis.

Monte Carlo Simulation
This book does not review the technique in detail but attempts to provide an overview of the principles behind Monte Carlo Simulation and highlight possible uses. The subject is particularly technical in its approach and a level of statistical understanding is required in order to fully appreciate the techniques used.

Monte Carlo Simulation is a statistical technique whereby randomly-generated data is used within predetermined parameters to simulate and produce realistic project outcomes. The overall project outcome is predicted by randomly simulating a combination of values for each risk and repeating the calculation, often up to 1000 times. Due to the nature of the calculation it is primarily a computer based operation. However, it is important to realise that the parameters (normally optimistic, realistic and pessimistic) and the appropriate distribution within which this random data is simulated is itself a series of subjective inputs. Accurate or realistic project outcomes will not be generated if inaccurate parameters are set.

It should be appreciated that where computerised systems are adopted in the quantitative assessment of risk there is still a necessity to understand the statistical techniques involved. It is essential that computer derived information is accompanied by written and well-documented supporting evidence scrutinising how the results have been achieved including a considered assessment of their viability and the methodology utilised therein.

Monte Carlo Simulation is a method of experimenting using a model within a set framework but containing random factors. Monte Carlo Simulation can be used to generate costs or rates within a defined range based upon the analysis of historical data, e.g. element unit rates for substructures may range from £35/m² to £75/m² plan area. At either end of the range the frequency (i.e. the number of times the event occurs) will be far lower than the middle. An example of this would be rolling two dice say 50 times. The result could be:

Outcome	Frequency (Number of occurrences)
2	1
3	1
4	3
5	5
6	9
7	13
8	10
9	4
10	2
11	2
12	0
	50 = total number of occurrences

Having established such a range (statistically known as a distribution) from historical data, a computerised random number selector can be used to generate random samples within that range weighted towards the mean (i.e. the number which has the highest probability of occurrence) which in our dice example is 7.

Take five basic elements:

>Substructure
>Superstructure
>Finishings
>Fixtures and fittings
>Mechanical and electrical installations

Each random number generated for each element is added together to complete one simulation as shown below, e.g.

Element	Simulation 1	Simulation 2	Simulation 3 etc.
Substructure	10,000	12,000	12,000
Superstructure	100,000	104,000	106,000
Finishings	20,000	19,00	21,000
Fixtures & fittings	20,000	20,000	19,000
M & E installations	60,000	75,000	67,000
TOTALS	£210,000	230,000	225,000

If this were completed 1000 times then the results could be plotted and analysed in a similar way to the dice exercise above and the statistical mean calculated. This method of simulation can be used to obtain an average risk estimate and a maximum likely risk estimate by applying confidence limits to the data produced.

Advantages of Monte Carlo Simulation

- a well thought out model for simulation should clearly quantify the risks and enable confidence limits to be set
- enables risks to be considered and experienced outside the real life of the project.

Disadvantages of Monte Carlo Simulation

- generally requires a dedicated computer software package and is therefore costly to run
- requires a knowledge of statistical techniques.

Latin Hypercube Sampling

As a further enhancement to simulation techniques Latin Hypercube Sampling has recently been developed as a method of streamlining Monte Carlo Simulations. In a normal Monte Carlo Simulation a typical number of individual runs required to achieve convergence (i.e. to be reasonably sure that the mean result is realistic) is in the order of 1000. Such a large number of runs is required to ensure that all variables are suitably sampled and where the model is complex it may take several days to complete. Latin Hypercube Sampling is a system whereby the sampling process is undertaken in a systematic rather than random manner, the result being that the number of runs required will be reduced from 1000 to something in the order of sixtyfour.

7

The risk management regime

Introduction

Effective risk management relies upon the formulation and implementation of a pre-determined regime governing reporting, presentation and responsibilities all as defined by the project manager in his RM Strategy.

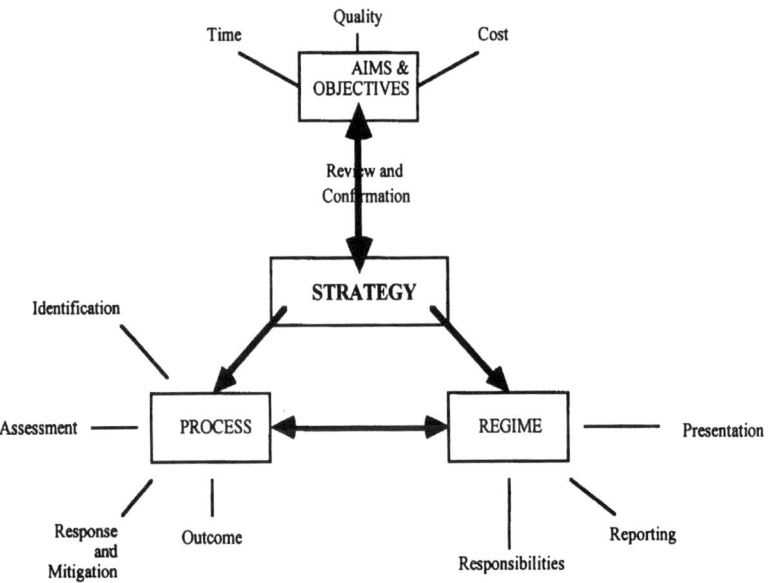

Figure 30. *Risk management regime*

In terms of risk responsibility, the aim is to formulate procedures defining and clearly demonstrating the flow of information across and between project participants, whilst appreciating the various degrees of confidentiality that may exist therein.

The responsibility for the overall management of risk lies firmly with the risk manager who will either be the project manager or his designated risk manager. By definition he is responsible for the implementation of a suitable regime and compliance therewith.

Under the direction of the risk manager individual risks must be managed by the most appropriate person or party who is empowered to take the management action necessary to mitigate its potential impact. Generally, however, heads of risks can be allocated to certain individuals or bodies as indicated in the two figures on the following pages.

Reporting

This section deals with the following topics:

- why report? – the answer may seem obvious but it is not just letting other people know what is happening.
- who in a risk management (RM) regime is responsible for reporting and where does the buck stop?
- what is reported?

Reporting is part of the communication process, it can be formal or informal, the process is both upwards and downwards in the chain of responsibility and applies to all stages of the project from inception to completion. It is the passing of information from one party to another to facilitate informed decision-making.

The criteria for reporting under a RM regime is no different to any other project management reporting function. Effective RM, however, relies upon the formulation and implementation of a pre-determined regime governing reporting, presentation and responsibilities.

This is not an attempt to describe a prescribed standard

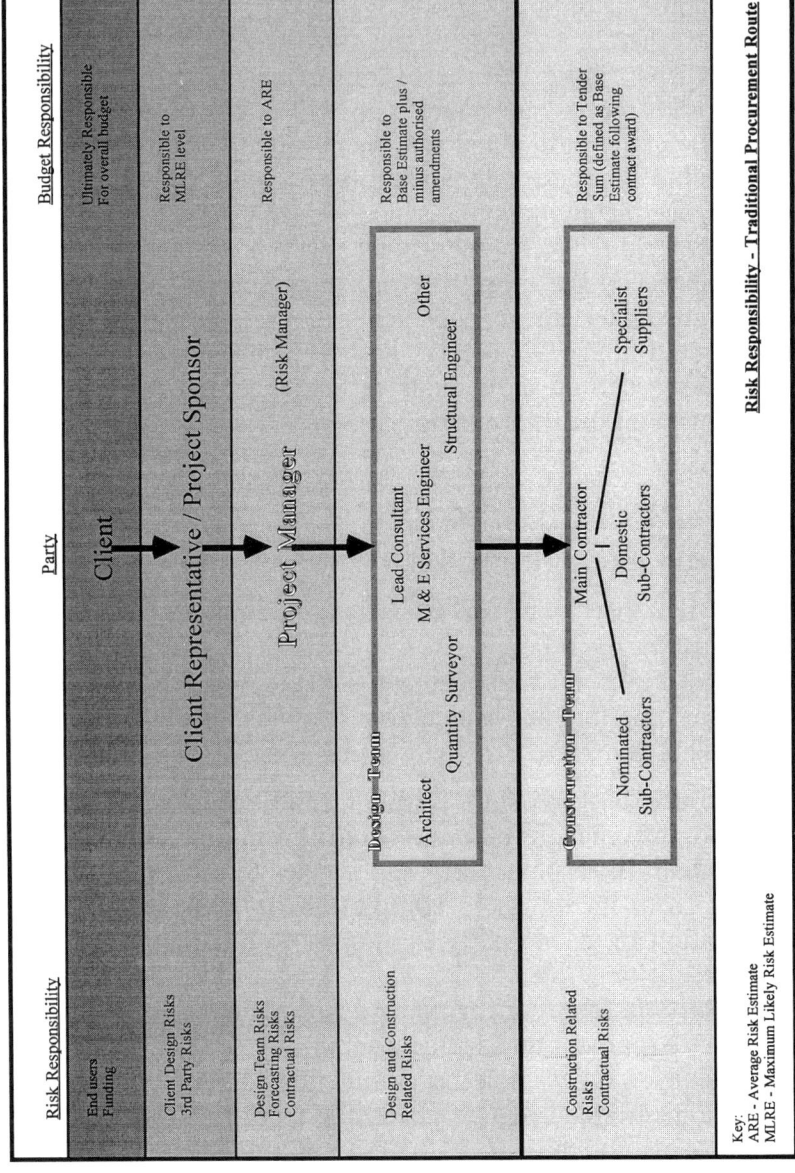

Figure 31. *Risk responsibilities*

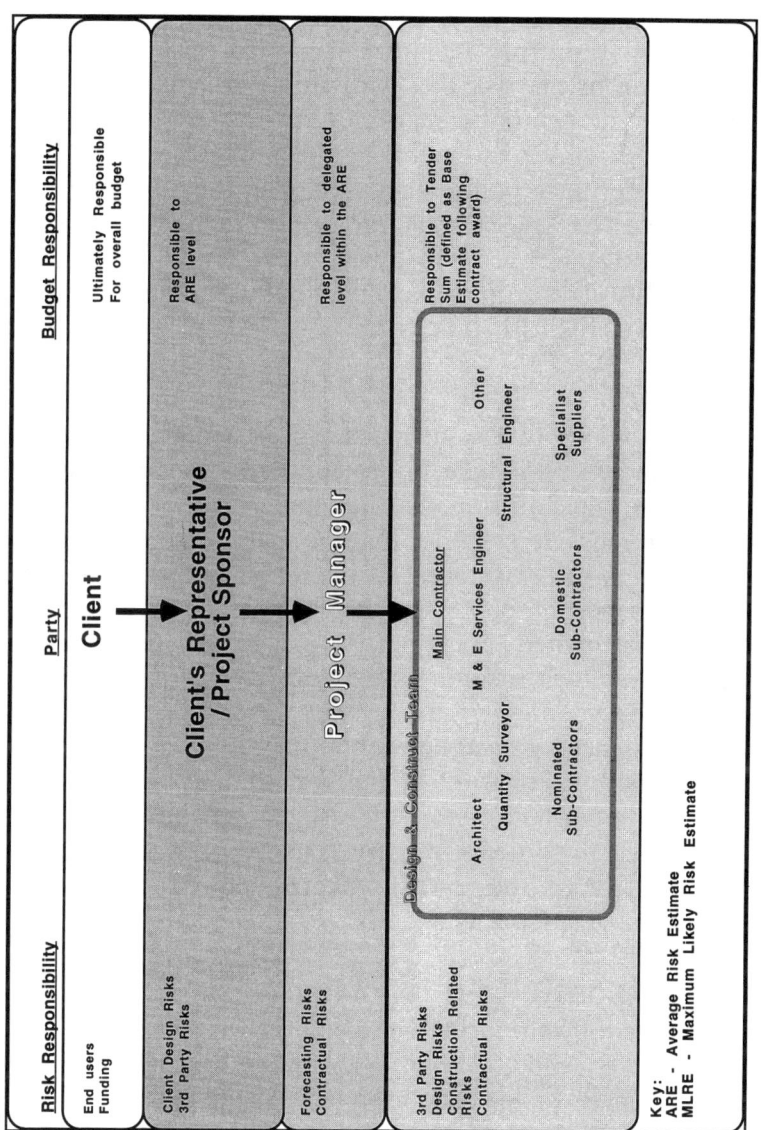

Figure 32. *Risk and budget responsibilities (design and build)*

reporting system but a suggestion of how to utilise the information generated by the risk-management process. Clients and projects vary, following the criteria of good practice will therefore be more likely to lead to greater understanding of the risk than will filling in standard forms regardless of the appropriateness.

In terms of risk responsibility the aim is to formulate procedures defining and clearly demonstrating the flow of information across and between projects participants, whilst at the same time appreciating that there may exist an element of confidentiality. The project manager would set out in advance such matters in his RM strategy.

Why report?

It may be obvious that some reporting procedures would be necessary in order to share information and seek instruction but RM reporting fulfils another function.

The main tangible evidence that the management of risk is taking place is the existence and implementation of the pre-determined reporting procedure. It is the means by which the process can be audited - compliance is measurable through each of the stages of identification, assessment, response, mitigation and outcome.

The experienced gained on each project is a valuable asset. The analysis and collation of the comprehensive reporting records over a period of time will result in much relevant and useful data and the process of analysis itself will lead to greater understanding of the various factors and incremental improvements to each RM undertaken.

Who is responsible for reporting?

Projects develop from inception to completion and during this progression, the responsibility for the individual risks will fall to differing parties. Indeed, prior to the appointment of the project manager, the client, through his project sponsor, will carry the burden of the overall RM regime. Appointment of the project manager and later in the process, the appointment of the contractor as key milestones in the responsibility of risk

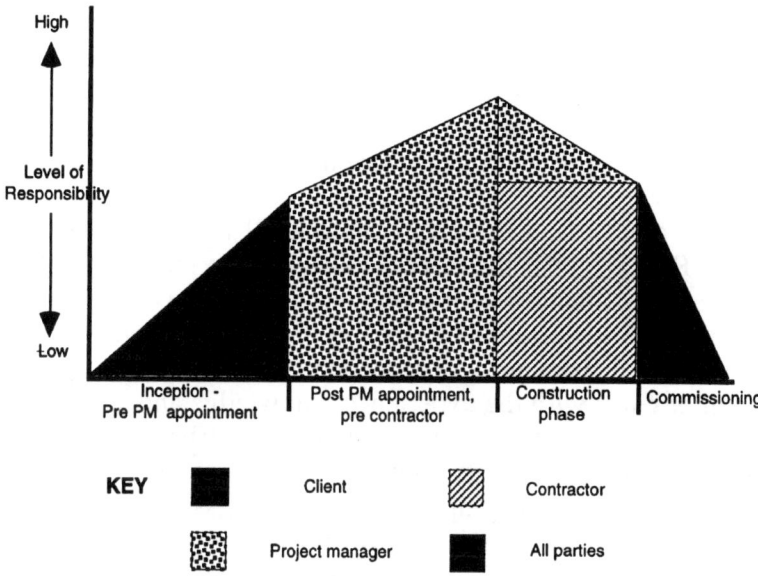

Figure 33. *Pre-dominant risk management responsibilities*

management. The pre-dominant risk management responsibilities can be illustrated as below

Once appointed, it is always the project manager's responsibility to ensure that the communication process of reporting is carried out efficiently, i.e. it facilitates the successful conclusion of the project. It is also the client's responsibility to appoint a project sponsor from within the organisation. It is important that the client organisation identifies from within its own ranks, a person who all recognise as being the focus for the particular project.

Project management is the process whereby the outcome of a project is achieved against pre-determined criteria. The definition of a project is an event which has a defined start and completion and in this sense differs from conventional management. The project manager controls the process which, in construction, is usually against the criteria of time, quality and cost. Project management is not necessarily the appointment of a single person – large and complex projects will involve a team.

The process can be thought of as a conduit, into one end goes the chaos and out of the other emerges a successful conclusion. What sometimes cannot be seen is the application of order, logic, discipline and experience going on inside the tube. Risk analysis and risk management is a fundamental part of this process.

The project manager carries the overall responsibility and that, in RM, includes the responsibility for the management of risk, the implementation of a suitable regime and compliance.

Whilst the project manager carries the overall responsibility, the individual risks must be managed by the most appropriate person or party with the experience, knowledge and skills best suited to the risk in question. Authority must be delegated to enable the person or party to mitigate the potential impact of the risk that they managing. The individual risk managers will follow all the principles laid down for the project as a whole including, of course, all of the reporting criteria. A delegation tree will emerge which may resemble Figure 34.

Figure 34. *Delegation*

A project manager will have delegated authority from the project sponsor, the terms of this authority need to be spelt out unambiguously. It follows that when the delegated powers are insufficient, he must seek client's instructions to enable progress to continue. Knowing exactly what can be dealt with by delegated powers and what the project manager must seek approval for, is important and it is equally important that the same disciplines are applied by the project manager to the individual risk managers.

An examination of the individual responsibilities indicates that whilst they follow the broad pattern of the pre-dominant risk management responsibilities, they are influenced by the timing and content of the project stages. Prior to the appointment of the project manager , the clients project sponsor bears the total responsibility for the risk. Once the project manager has been appointed, the project sponsor hands over much of that responsibility and, following briefing and agreement of the project manager's scope of service, the sponsor's responsibility declines as the project manager's increases.

The appointment of the contractor begins his responsibility for the construction phase and marks the point at which construction-related risk factors are mitigated by contractual arrangements. The project manager still has to ensure that all outstanding matters are resolved and the contractor instructed. Commissioning draws the project to a conclusion with the contractor still assuming an on-going responsibility for defects. The individual responsibility is illustrated in Figure 35.

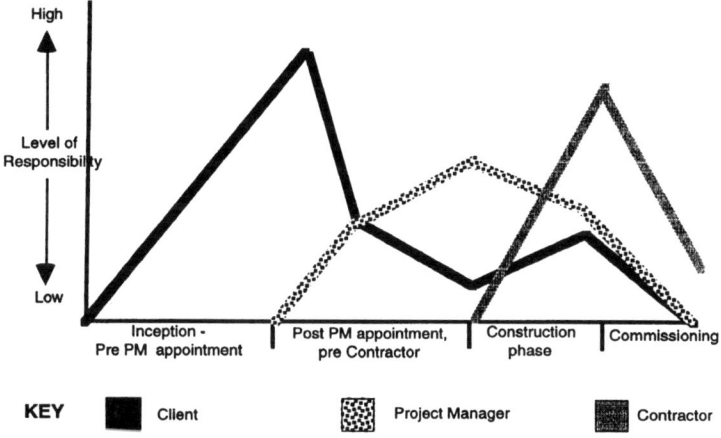

Figure 35. *Individual risk responsibilities*

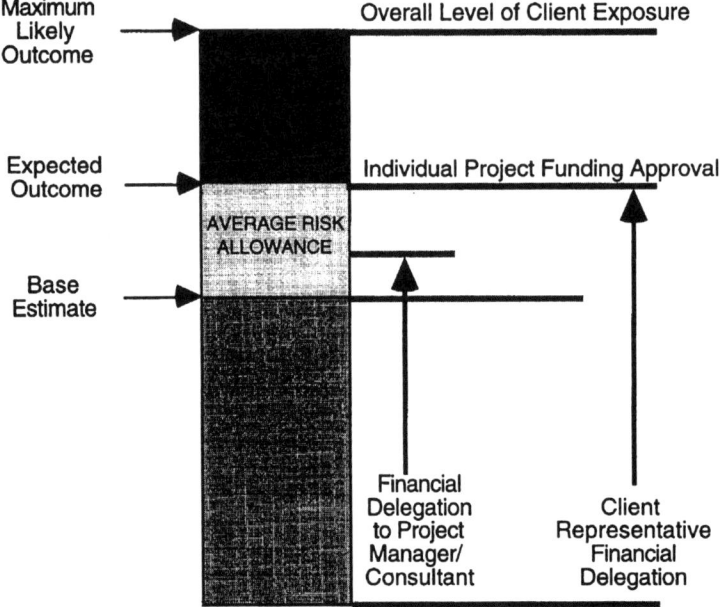

Figure 36. *Typical budgetary responsibilities*

Finally, under the heading of responsibilities, is the question of financial responsibility. Client project approval should be at the maximum likely risk estimate. It is recommended that project sponsor is authorised up to the average risk estimate and the project manager between that and the base estimate level. The degree of the project manager's authorisation over and above the base estimate is a matter influenced by particular circumstances. Figure 36 illustrates these recommendations.

What is reporting?

The report is one of the project manager's tools of the trade. It is one of the means by which he obtains decisions and initiates actions – actions that result in a successful project.

In a RM regime, reporting can be considered as the essential outcome of continual monitoring, review and assessment/reassessment of risk. The reporting should be an appropriate, concise and accurate procedure.

When considering the reporting under a risk management

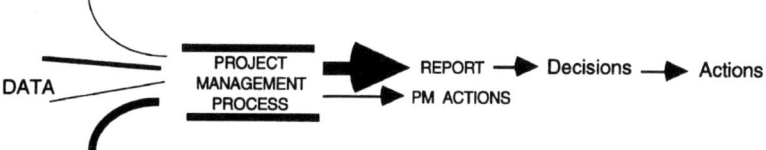

Figure 37. *The project management process*

regime (RMR), it may be helpful to consider a few basics. The information contained in a report consists of a collection of data assembled and influenced by the project management process.

How and when?

Basic criteria
There are certain basic criteria that apply to all reports especially those which deal with complex issues. Reports should follow four golden rules. They should be:

- clear and well presented
- relevant
- timely
- reflect authority.

Clarity and presentation
There is nothing more off-putting than receiving a report which is difficult to follow or is set out in a confusing and illogical manner. It must be a prime objective of the report-writer to assist the recipient by presenting a report in such a format that it assists in the process of assimilating the information contained in the report. Executive time is at a premium, especially those executives which will be in the position to make the decisions required of a report. Consideration should therefore be given to the technique of layering the report into levels of detail. For example:

LEVEL 1   Executive summary aimed at board level consideration.

LEVEL 2   Detailed summary with sufficient detailed for senior management.

LEVEL 3   Supporting detail to enable technical appreciation of principal factors.

Technical content can be summarised as below:

Figure 38. *Reporting levels*

Risk will probably be well understood in a client's organisation but this will be the entrepreneurial type of risk associated with the company's operations. The client may not understand the risks associated with a major construction project. Clear reporting with adequate supporting details will help that understanding and serve also to illustrate, for those who need to know, the report's background and influencing factors. Clear concise reporting along with the monitoring of risks through the risk register is most important. A typical risk register can be seen in Figure 39.

Technical jargon should be avoided especially in the summary sections. Where possible relate the report to the client's key criteria. For example, financial risk could be related to the client's financial year or time risk to peak trading periods.

Ref.	Brief description	Trigger Event	Interdependencies (Ref to)	Pb	Effect	Impact ARA MLRA	Response/mitigation Status Current Previous	Owner	Status	Comments/ actions

T = Third party risks
S = Site risks
Cl = Client risks
D = Design team risks
Co = Contractors risks

*Ref to is the reference to the dependent risk(s)

T = Time
C = Cost
Q = Quality

A = Assessed
M = Managed out
D = Designed out
S = Shared/transferred
I = Ignored

I =- Initial/intuitive
C = Considered
F = Final

Figure 39. *Risk register*

INTERIM PROJECT RISK REPORT

Project

Report Number _____   **Date** _____

Initial / Previous Estimate		**Current Estimate**	
Base Estimate	[ ]	Confirmed Costs	[ ]
Risk Allowance	[ ]	Revised Risk Allowance	[ ]
Expected Outcome	[ ]	Anticipated Out-turn Costs	[ ]
Target Completion Date	[ ]	Contract Completion	[ ]
Risk Allowance	[ ]	Approved Extensions	[ ]
Expected Completion	[ ]	Risk Allowance	[ ]
		Anticipated Completion	[ ]

Summary of Risks Still to Be Expended	Average Allowance	Maximum Allowance

Notes:

Signed _____

For _____

Figure 40. *Typical interim risk report*

Relevance

Relevance not only relates to the obvious need for a report to be relevant to the task at hand, but the need to be relevant to the company culture of the client or project sponsor. Examples of relevance or irrelevance can include:

- able to be clearly understood by intended recipients
- attention to too much detail when fundamentals are unanswered
- matching the report to the decision-making process of the client
- being available at the right time
- transmitting to all users and preferable all providers of the information.

Risk impacts on the client's company or organisations results, it could impact on sales, customer satisfaction or even survival. The reporting process is the means by which the company is kept informed of progress, matters on which it has to decide and, insofar as clarity is concerned, it should be presented in terms that the client understands and can relate to. The stakes are high and the project manager cannot hide behind jargon and techno-mumbo-jumbo.

Timing

Late information is nearly as bad as no information. It is easy to focus on critical project dates relating to site acquisition, planning permission and or construction phases, but the client's organisation will have critical dates also. Financial year-ends, formal board meetings, cash flow, market or trading opportunities are examples of a few. Even executives' holidays can be a critical factor when progress requires their decisions. It is important to match the reporting (and the need to obtain decisions) with the client's critical dates with appropriate lead-in time to allow consideration and distribution within the client's organisation.

Authority

On a wider but related issue is the whole question of the client's company culture. All organisations differ and have their own unique way of approaching decision-making, especially when it relates to a major capital expenditure. The terms of the project manager's appointment will or should reflect this. This is for the benefit of both the client and the project manager – misunderstandings in a risk-management regime can cost money. It follows that the report will consist of both matters the project manager has actioned under his own authority and those matters that require the client's decision.

Contingency management

Having adopted risk management, contingency levels should be a clear reflection of the project's risk exposure. That being the case, expenditure against these residual risk items must be controlled (clearly there will be a small element of contingency for uncontrollable issues). The risk allowances will all be calculated from the detailed quantitative assessment (risk analysis) undertaken previously.

As indicated in Figure 36, Budget Responsibilites contingencies are generally set within three main areas:

- *construction contingency*   this would normally be an allowance to reflect those risks considered to be construction risks and within the control of that team. It will be at a level designated by the client's representative to the appropriate construction team leader
- *design contingency*   this is an allowance mainly for use during the design process and will normally take in all risks up to the expected out-turn costs or Average Risk Estimate. It is usually managed by the client's representative or the project manager
- *client contingency*   this is the allowance held by the client to reflect the risk of client changes. It takes in all risk exposure up to the Maximum Likely Risk Estimate level. The client will hold a number of such contingencies and may have the ability to transfer funds across projects if necessary.

It is important that any contingencies are set realistically to reflect the status of the project. Contingencies set too highly to avoid future embarrassment can be as detrimental to the client as those set too low in that funds may be incorrectly diverted or even the project may be delayed or aborted.

In order to support good contingency management, the risk management regime must place facilities to enforce:

- change control
- control of as-yet unmaterialised risk
- time, cost and quality control.

Clearly, the long term aim of contingency management within the control of a risk management regime is to ensure that projects are delivered at the lowest cost compatible with the specified quality and in the desired time. Not only this, but for a project to be successful it is essential that the project outcome is the same or very similar to the expected outcome when approval was granted.

8
Information technology

Introduction

Risk analysis is a growth area and there are now several software programmes available, particularly for modelling and simulation. The majority of these are for use on IBM compatible PCs but other formats are available including Apple Macintosh. However, the majority of techniques outlined in this chapter can be accommodated by off the shelf spreadsheet packages such as Excel and Lotus 123, the exception generally being Monte Carlo Simulation.

Most organisations are now readily equipped with the latest information technology (IT) and in particular the spreadsheet has become an invaluable tool. The use of such technology should allow more time to be available for other activities thus enabling the overall quality of work to increase. Inevitably, the result has been to bring pressure for quicker results but benefits have arisen and in particular probably the greatest advantage has been the ability to do 'what-ifs'. By quickly changing a couple of figures in a spreadsheet we can see the results immediately. One of the areas where this has been of greatest benefit in the construction industry has been in undertaking investment appraisals or feasibility studies. Here a small change in interest rates or yield can dramatically effect the profitably of any project.

The following pages show spreadsheet examples of investment appraisals and feasibility studies where spreadsheets have been used to sensitivity-test the overall scheme.

DEVELOPMENT SUMMARY

Financial Information   **Comments**

Deposit Rate: Interest (%p.a.)	4.5
Cost of Capital:Interest(% p.a.)	9.0
Investment Value	£5,987,692
Contingency (percentage of costs)	

Total value of houses plus commercial
Some contingency in construction costs

Time:

Time lag to building start	2 Months
Construction period	36 Months
Month of Sales start	18 Months
Sales lag after completion	4 Months
Development period therefore =	42 Months

assuming land bought month 0,
construction start month 4

Summary

LAND	765,000
CONSTRUCTION COSTS	4,223,000
SOLICITORS LEGAL & EST. AGENTS	3,636
NHBC& BUILDING REG/PLNG FEES	38,000
HOUSE SALES FEES	76,000
PROFESSIONAL FEES	191,150
INTEREST ON CAPITAL (see cash flow)	incl
CONTINGENCIES	
DEVELOPMENT COSTS	£5,296,786
DEVELOPERS MARGIN	£916,902
MARGIN AS % OF COST	17.31%

Interest is calculated within cash flow

Balance from cash flow at end of period

Figure 41: *Typical development appraisal*

Development Values:

Site:

Area (acres)	Density	Mix:	Over-ride
		Type A	100%
% useable	Net Area	B	
for housing		C	
	Total Units 38 units	D	

Unit Data

Unit type	Nr of Units	Area	Cost/ ft²	Build Cost	Sale Value	Total Sales
A	38	1300 ft²	£45.00	58,500	125,000	4,750,000
B		65 ft²				
C		65 ft²				
D		65 ft²				

Total Units <u>38</u>   Average build cost <u>58,500</u>

Average sell price <u>125,000</u>

TOTAL HOUSING
CONSTRUCTION COSTS   <u>£2,223,000</u>

              TOTAL SALE VALUE   <u>£4,750,000</u>

Commercial
Public House
    SALE OF LAND FOR A CHANDLERY: <u>150,000</u>
Marina Berths
No of Berths   130
Water Frontage 28 ft

Lease per ft	£60.00
Brokerage and craneage p.a.	50,000
Maintenance and dredging costs p.a.	127,000
Total Per annum	141,400
Capitalised at a yield of	13 00%
CAPITAL VALUE:	<u>£9087,592</u>
TOTAL CAPITAL VALUE	<u>£1,237,002</u>

Figure 42: *Typical development appraisal*

Development Costs

A ACQUISITIONS	%	Cost
Land	750,000	
Legal Fees on Land Purchase	1	7,500
Stamp Duty	1	7,500
Other Land Fees		
TOTAL LAND PURCHASE COSTS		765,000

B CONSTRUCTION	Cost	Start Month	Duration	Rate p.a.
Marina Berths		1,250,000	1	12
General infrastructure	750,000	1	12	
Chandlery		30	1	
Flats 0 Nr				
Houses 38Nr @ £58,500	2,223,000	10	26	18
			4	
TOTAL CONSTRUCTION	4,223,000			

C FEES	%	Fee	Override	% upfront
Construction related Fees:				
Employers Agent (% of Con St.)	2%	84,460		
Engineer (% of Civils)	2%	40,000		
Architect (% of bids)	3%	66,690		
Legals/Agent's Fees:				
Berths/Retail Premises (% of val.)		30,000	30,000	
House sales (Fee per unit)	2,000	76,000		
Other Fees:				
NHBC/bldgreg per unit	1000	38,000		
Any other fees				
TOTAL FEES		335,150		

D CONTINGENCY (incl)

DEVELOPMENT COSTS EX. INTEREST	5,323,150
MAXIMUM CAPITAL EMPLOYED	3,348,263

Figure 43. *Typical development appraisal*

COSTS									
MTH	LAND	BERTHS	INFRA	FLATS	HOUSES	BUILD COST	EMPLOYER'S AGENT	ENGINEER	ARCHITECT
1	765,000								
2									
3		24,115	14,469			38,584	772	772	
4		25,219	15,131			40,350	807	807	
5		43,173	25,904			69,077	1,382	1,382	
6		71,533	42,920			114,453	2,289	2,289	
7		104,978	62,987			167,964	3,359	3,359	
8		138,185	82,911			221,097	4,422	4,422	
9		165,836	99,501			265,337	5,307	5,307	
10		182,608	109,565			292,172	5,843	5,843	
11		183,181	109,908			293,089	5,862	5,862	
12		162,233	97,340		22,543	282,116	5,642	5,191	676
13		114,445	68,667		17,476	200,588	4,012	3,662	524
14		34,495	20,697		17,9117	3,102	1,462	1,104	537
15					21,417	21,417	428		643
16					27,566	27,566	551		827
17					35,927	35,927	719		1,078
18					46,073	46,073	921		1,382
19					57,572	57,572	1,151		1,727
20					69,996	69,996	1,400		2,100
21					82,915	82,915	1,658	2,487	
22					95,900	95,900	1,918		2,877
23					108,522	108,522	2,170		3,256
24					120,352	120,352	2,407		3,611
25					130,958	130,958	2,619		3,929
26					139,914	139,914	2,798		4,197
27					146,788	146,788	2,936		4,404
28					151,153	151,153	3,023		4,535
29					152,577	152,577	3,052		4,577
30					150,633	150,633	3,013		4,519
31					144,890	144,890	2,898		4,347
32					134,919	134,919	2,698		4,048
33					120,291	120,291	2,406		3,609
34					100,576	100,576	2,012		3,017
35					75,345	75,345	1,507		2,260
36					44,169	44,169	883		1,325
37					6,618	6,618	132		199
38									
39									
40									
41									
42									
TOTALS	£1,250,000	750,000			2,223,000	4,223,000	84,460	40,000	66,690

Figure 44. *Typical development appraisal*

INCOME

CONST FEES	COMM FEES	LHOUSE FEES	NHBC BLD.REGS.	TOTAL (/MONTH)	TOTAL (CUM.)	BERTH RENT	BERTHS	LAND SALE	DEPOSIT INTEREST
				765,000	765,000				
					765,000				
					765,000				
1,543				40,127	805,127				
1,614				41,964	847,091				27,763
2,763				71,840	918,931				
4,578				119,031	1,037,963				
6,719				174,683	1,212,646				24,608
8,844				229,940	1,442,586				
10,613				275,951	1,718,537				
11,687				303,859	2,022,396				16,381
11,724				304,812	2,327,208				
11,510			38,000	331,626	2,658,835	80,000			
8,198				208,786	2,867,621				6,039
3,103				76,206	2,943,827				
1,071				22,488	2,966,315				
1,378				28,944	2,995,259				1,801
1,796				37,724	3,032,983				
2,304		2,000		50,376	3,083,359				
2,879		2,000		62,450	3,145,809				1,063
3,500		2,000		75,496	3,221,305				
4,146		2,000		89,061	3,310,366				
4,795		2,000		102,695	3,413,061				1,892
5,426		2,000		115,948	3,529,009				
6,018		2,000		128,369	3,657,379	80,000			
6,548		2,000		139,506	3,796,885				2,281
6,996		2,000		148,910	3,945,795				
7,339		2,000		156,128	4,101,922				
7,558		2,000		160,710	4,262,633				677
7,629		4,000		164,206	4,426,839				
7,532	3,636	4,000		165,800	4,592,639			150,000	
7,244		4,000		156,134	4,748,773				3,035
6,746		4,000		145,665	4,894,437				
6,015		4,000		130,305	5,024,743				
5,029		4,000		109,605	5,134,347				5,798
3,767		4,000		83,112	5,217,459				
2,208		4,000		50,377	5,267,837	150,000			
331		4,000		10,949	5,278,786				12,406
		4,000		4,000	5,282,786				
		4,000		4,000	5,286,786				
		4,000		4,000	5,290,786				19,922
		2,000		2,000	5,292,786				
		2,000		2,000	5,294,786				
191,150	3,636	100,000	38,000	5,320,786					

Figure 45. *Typical development appraisal*

			HOUSE SALES		CASH FLOW		
INTEREST/ COMMERCIAL INCOME	CUMULATIVE	RATE OF COMPLETION	SELLING PRICE	TOTAL (CUM)	INTEREST	INTEREST CUMULATIVE	CASHFLOW
							(765,000)
					25,112	25,112	(790,112)
					25,112	50,224	(815,224)
					25,112	75,336	(880,463)
27,763	27,763				25,112	100,448	(919,776)
	27,763				25,112	125,560	(1,016,729)
	27,763				25,112	150,672	(1,160,872)
24,608	52,371				25,112	175,784	(1,336,058)
	52,371				25,112	200,896	(1,591,111)
	52,371				25,112	226,008	(1,892,174)
16,381	68,752				25,112	251,120	(2,204,764)
	68,752				25,112	276,232	(2,534,688)
80,000	148,752				25,112	301,344	(2,811,427)
6,039	154,791				25,112	326,456	(3,039,285)
	154,791				25,112	351,568	(3,140,603)
	154,791				25,112	376,680	(3,188,203)
1,801	156,592				25,112	401,792	(3,240,459)
	156,592				25,112	426,904	(3,303,294)
	156,592	1	125,000	125,000	25,112	452,016	(3,253,782)
1,063	157,655	1	125,000	250,000	25,112	477,128	(3,215,282)
	157,655	1	125,000	375,000	25,112	502,240	(3,190,889)
	157,655	1	125,000	500,000	25,112	527,351	(3,180,062)
1,892	159,547	1	125,000	625,000	25,112	552,463	(3,180,977)
	159,547	1	125,000	750,000	25,112	577,575	(3,197,038)
80,000	239,547	1	125,000	875,000	25,112	602,687	(3,145,519)
2,281	241,828	1	125,000	1,000,000	25,112	627,799	(3,182,856)
	241,828	1	125,000	1,125,000	25,112	652,911	(3,231,878)
	241,828	1	125,000	1,250,000	25,112	678,023	(3,288,118)
677	242,505	1	125,000	1,375,000	25,112	703,135	(3,348,263)
	242,505	2	250,000	1,625,000	25,112	728,247	(3,287,581)
150,000	392,505	2	250,000	1,875,000	25,112	753,359	(3,078,493)
3,035	395,540	2	250,000	2,125,000	25,112	778,471	(3,006,704)
	395,540	2	250,000	2,375,000	25,112	803,583	(2,927,481)
	395,540	2	250,000	2,625,000	25,112	828,695	(2,832,898)
5,798	401,337	2	250,000	2,875,000	25,112	853,807	(2,711,817)
	401,337	2	250,000	3,125,000	25,112	878,919	(2,570,041)
150,000	551,337	2	250,000	3,375,000	25,112	904,031	(2,245,530)
12,406	563,743	2	250,000	3,625,000	25,112	929,143	(2,019,186)
	563,743	2	250,000	3,875,000	25,112	954,255	(1,798,298)
	563,743	2	250,000	4,125,000	25,112	979,367	(1,577,410)
19,922	583,665	2	250,000	4,375,000	25,112	1,004,479	(1,336,599)
	583,665	1	125,000	4,500,000	25,112	1,029,591	(1,238,711)
	583,665	1	125,000	4,625,000	25,112	1,054,703	(1,140,823)
1,042,391		50	6,250,000		1,054,703		916,902

Figure 46. *Typical development appraisal*

The example on the preceding pages shows a development margin of 17.31% as a percentage of costs for a marina and housing development. If the developer suddenly has to pay a higher rate of interest on his borrowings of say 12%, then we can quickly calaculate what the effect would be on the overall profitability of the scheme.

DEVELOPMENT SUMMARY

Financial Information   **Comments**

Deposit Rate: Interest (% p.a.)	4.5	
Cost of Capital: Interest (% p.a)	12.0	
Investment Value	£5,987,692	Total value of houses plus commercial
Contingency (percentage of costs)		Some contingency in construction costs

Time:

Time lag to building start	2 Months	assuming land bought month 0,
Construction period	36 Months	construction start month 4
Month of Sales start	18 Months	
Sales lag after completion		4 Months
Development period therefore =		42 Months

Summary

LAND	765,0001	
CONSTRUCTION COSTS	4,223,000	
SOLICITORS LEGAL & EST. AGENTS	3,636	
NHBC & BUILDING REG/PLNG FEES	38,000	
HOUSE SALES FEES	76,000	
PROFESSIONAL FEES	191,150	
INTEREST ON CAPITAL (see cash flow)	incl	Interest is calculated within cash flow
CONTINGENCIES		
DEVELOPMENT COSTS	£5,296,786	
DEVELOPERS MARGIN	£468,202	Balance from cash flow at end of period
MARGIN AS % OF COST	**8.84%**	

Figure 47. *Development summary*

Now the margin is only 8.84% of costs.

As well as for feasibility studies for private developers, the spreadsheet can be used for sensitivity-analysis for longer-term investment appraisals of say 30 years, where you are trying to calculate net present values (NPV) or internal rates of return (IRR). An example of this can be seen on the following pages.

Unit Information		
Nr of units		50
Nr of persons accommodated		125
Nr of Bedspaces		75
Financial Information		
Discount Rate (Treasury rate)		6%
Inflation Rate		2%
Sinking Fund Rate		5%
Rental Growth		
Review Periods		0
Capital Costs		
Acquisitions		
Construction Works (incl demos)		
Refurbishment		£1,832,383
Demolitions		
Decanting/Rehousing		
Fees		£366,477
TOTAL		£2,198,860
Revenue Costs		
Management & Maintenance		£53,800
Annual Sinking Fund		£66,797
Service Costs		£36,300
Bad Debts		£15,000
Insurance		£10,000
Other		£3,000
Income		
Gross Rental Income		£91,000
Voids	10%	(9,100)
Net Rental		£81,900
Service Income		£39,000
RTB Sales		
Market Sales		0
Other		£1,500

Figure 48. *Typical NPV calculation*

The above table is used to produce a discounted cashflow for a large housing development, with resultant NPVs as follows:

	Year 0	Year 1	Year 2	Year 60	Total
CAPITAL COSTS					
Acquisitions					
Construction Works	2,198,860				2,198,860
Decanting/Rehousing					
Fees					
TOTAL	2,1 98,860				2,198,860
REVENUE COSTS					
Management					
Day to day maintenance	53,800	53,800	53,800	53,800	3,228,000
Sinking Fund	66,797	66,7976	6,797	66,797	4,007,836
Service Costs	36,300	36,300	36,300	36,300	2,178,000
Void Rent / Bad Debts	15,000	15,000	15,000	15,000	900,000
Insurance	10,000	10,000	10,000	10,000	600,000
Other	3,000	3,000	3,000	3,000	180,000
TOTAL		184,897	184,897	184,897	11,093,836
TOTAL COSTS					
(A1 + A2)	2,198,860	184,897	184,897	184,897	13,292,695
INCOME					
Rental Income	81,900	81,900	81,900	81,900	4,914,000
Service Income	39,000	39,000	39,000	39,000	2,340,000
Right to Buy Sales					
Market Sales	0	0	0	0	
Residuals (Year 60 only)					
Other	1,500	1,500	1,500	1,500	90,000
TOTAL		122,400	122,400	122,400	7,344,000
INVESTMENT VALUE	(2,198,860)	(62,497)	(62,497)	(62,497)	(5,948,695)
Other Notional Income	50,000	50,000	50,000	50,000	3,000,000
TOTAL		50,000	50,000	50,000	3,000,000
NET PROJECT VALUE	(2,198,860)	(62,497)	(62,497)	(62,497)	(5,948,695)
DISCOUNT FACTOR		1.0000 0.9434	0.8900	0.0303	
NET PRESENT VALUE	(2,198,860)	(58,960)	(55,622)	(1,895)	(3,208,905)

Figure 49. *Typical NPV calculation*

Net Present Value	(3,208,905)
NPV per Unit	(64,178)
NPV per person	(25,671)
N PV per bedspace	(42,785)

Figure 50. *NVP Outputs*

By using sensitivity testing we can analyse the results as shown below:

Rental Income

Variance in rental income from that included above	Variance	NPV	Per Unit
	50.00%	(1,712,360)	(34,247)
	25.00%	(2,043,265)	(40,865)
	5.00%	(2,307,989)	(46,160)
	-10.00%	(2,506,532)	(50,131)
	-20.00%	(2,638,895)	(52,778)
	-50.00%	(3,035,981)	(60,720)

Void Rate

Variance in void rate from that included above	Variance	NPV	Per Unit
	50.00%	(2,447,705)	(48,954)
	25.00%	(2,410,938)	(48,219)
	-10.00%	(2,388,877)	(47,778)
	-5.00%	(2,366,817)	(47,336)
	-10.00%	(2,359,463)	(47,189)
	-15.00%	(2,352,110)	(47,042)

Service Income

Variance in service charge from that included above	Variance	NPV	Per Unit
	10.00%	(2,311,141)	(46,223)
	5.00%	(2,342,656)	(46,853)
	-5.00%	(2,405,685)	(48,114)
	-10.00%	(2,437,200)	(48,744)
	-20.00%	(2,500,230)	(50,005)
	-30.00%	(2,563,259)	(51,265)

Figure 51. *Sensitivity Analysis*

Indirect Benefits

Variance in indirect benefits from that included above	Variance	NPV	Per Unit
	100.00%	(1,566,099)	(31,322)
	50.00%	(1,970,135)	(39,403)
	25.00%	(2,172,153)	(43,443)
	-50.00%	(2,778,206)	(55,564)
	-100.00%	(3,182,242)	(63,645)
	-200.00%	(3,182,242)	(63,645)

RTB Sales

Variance in RTB sales from that included above	Variance	NPV	Per Unit
	50.00%	(2,374,170)	(47,483)
	25.00%	(2,374,170)	(47,483)
	10.00%	(2,374,170)	(47,483)
	-5.00%	(2,374,170)	(47,483)
	-10.00%	(2,374,170)	(47,483)
	-15.00%	(2,374,170)	(47,483)

Residual Value

Variance in residual value from that included above	Variance	NPV	Per Unit
	50.00%	(2,360,839)	(47,217)
	25.00%	(2,367,505)	(47,350)
	-10.00%	(2,376,837)	(47,537)
	-50.00%	(2,387,502)	(47,750)
	-75.00%	(2,394,167)	(47,883)
	-100.00%	(2,400,833)	(48,017)

Figure 52. *Sensitivity analysis*

Maintenance & Management

Variance in Maintenance & Management from that included above	Variance	NPV	Per Unit
	25.00%	(2,618,204)	(52,364)
	10.00%	(2,487,782)	(49,756)
	5.00%	(2,444,307)	(48,886)
	-6.00%	(2,357,359)	(47,147)
	-10.00%	(2,313,885)	(46,278)
	-25.00%	(1,531,348)	(30,627)

Sinking Fund

Variance in sinking fund from that included above	Variance	NPV	Per Unit
	50.00%	(2,940,603)	(58,812)
	25.00%	(2,670,718)	(53,414)
	10.00%	(2,508,787)	(50,176)
	-10.00%	(2,292,879)	(45,858)
	-25.00%	(2,130,948)	(42,619)
	-50.00%	(1,861,064)	(37,221)

Service Costs

Variance in service costs Variance NPV Per Unit
fromthat included above

	50.00%	(2,694,163)	(53,883)
	25.00%	(2,547,498)	(50,950)
	10.00%	(2,459,499)	(49,190)
	-5.00%	(2,371,500)	(47,430)
	-10.00%	(2,342,167)	(46,843)
	-15.00%	(2,312,834)	(46,257)

Figure 53. *Sensitivity analysis*

Construction

Variance in construction fromthat included above	Variance	NPV	Per Unit
	20.00%	(2,767,310)	(55,346)
	10.00%	(2,584,071)	(51,681)
	5.00%	(2,492,452)	(49,849)
	-5.00%	(2,309,214)	(46,184)
	-10.00%	(2,217,595)	(44,352)
	-20.00%	(568,450)	(11,369)

Fees

Variance in Fees from that included above	Variance	NPV	Per Unit
	25.00%	(2,492,452)	(49,849)
	10.00%	(2,437,481)	(48,750)
	5.00%	(2,419,157)	(48,383)
	-5.00%	(2,382,509)	(47,650)
	- 10.00%	(2,364,185)	(47,284)
	-25.00%	(2,309,214)	(46,184)

Figure 54. *Sensitivity analysis*

The preceding tables all show how spreadsheets can be used for sensitivity-analysis purposes. In addition most of the other risk analysis techniques discussed earlier in this book can also be set up on standard spreadsheet software. The examples on the following pages are for use with Excel version 4.

Probabilistic estimating

A simple probabilistic estimate can be set up using the following formulas:

	F	G	H	I	J		K	L
	Optimistic	Pb.	Realistic	Pb.	Pessimistic		Pb.	Expected
	Cost		Cost		Cost			Value
7	25000	0.1	27000	0.5	30000		0.4	=(F7*G7)+(H7*17)+(J7*K7)
8								=(F8*G8)+(H8*I8)+(J8*K8)
9	15000	0.2	18000	0.6	21000		0.2	=(F9*G9)+(H9*19)+(J9*K9)
10	1500	0.2	2000	0.7	2500		0.1	=(F10*G10)+(H10*I10)+(J10*K10)
11	12000	0.3	15000	0.5	17500		0.2	=(F11*G11)+(H11*I11)+(J11*K11)

Figure 55. *Probabilistic estimates*

The numbers in the left hand column indicate the row numbers and the letters at the top are the colurnn references.

MERA (Multiple Estimating Using Risk Analysis)

Average Risk	Maximum	Spread	Spread Squared
20000	50000 = I8-H8	=J8*J8	
		=I9-H9	=J9*J9
50000	200000	=I10-H10	=J10*J10
		= I11-H11	= J11*J11
		=I12-H12	=J12*J12
30000	100000	=I13-H13	=J13*J13
		=I14-H14	=J14*J14
		=I15-H15	=J15*J15
		=I16-H16	=J16*J16
		=I17-H17	=J17*J17
		=I1 8-H18	=J18*J18
		=I19-H19	=J19*J19
		=I20-H20	=J20*J20
		=I21-H21	=J21 *J21
		=I22-H22	=J22*J22
		=I23-H23	=J23*J23
		=I24-H24	=J24*J24
=SUM(H8:H23)		To Summary	=SUM(K8:K23)

Figure 56. *MERA*

Here the same principle would apply, the colurnns are referenced alphabetically and the rows numerically. A calculation such as this would be undertaken for each category of risk the summed and square rooted as indicated below:

Average	Risk	Spread	Square
=thirdav		=thirdmax	
=siteav		=sitemax	
=clientav		=clientmax	
=consav		=consmax	
=contav		=contmax	
=otherav		=othermax	
=SUM(B6:B11)		=SUM(C6:C11)	
=C2	Average Risk Esl=B17		
= B 13	Square root of sp =SQ RT(C13)		
=SUM(B1 5:B1	Max. Likely I =SUM(pi 5:D16)		

Figure 57. *MERA*

By naming cells as indicated in the previous example (e.g. thirdav, clientav, sitemax) the calculation is made easier by avoiding disrupting the spreadsheet framework when new columns or rows are added.

Having given an insight into spreadsheets and how they can be used to set up simple risk analysis calculations, it is also worth considering what can be purchased off the shelf. A list of dedicated risk analysis software can be obtained from the Association of Project Managers but the following is a brief summary of those available.

Off the shelf risk analysis software
It is not the intention of this book to endorse any particular risk analysis software but merely to list some of the products that are available.

Software	*Format*
@RISK	available for both Apple Macintosh and DOS based computers as an add on to either Excel or Lotus 1-2-3; also

	available as an add in for Microsoft Project for DOS PCs only.
AS (Application System)	software package for use on IBM compatible PCs only
CASPAR	software package for use on IBM compatible PCs only (286 or greater)
CRYSTAL BALL	add on to Lotus 1-2-3 or Microsoft Excel on both IBM (Windows 3.1) and Apple Macintosh format.
DMT	software package for use on IBM compatible PCs only (386 or greater with 8Mb of RAM)
DynRisk	software programme that interfaces with Microsft Excel, Project and Word on Apple Macintosh system 7 or greater.
Monte Carlo	software that interfaces with dBASE on IBM compatible 386 with 4Mb of RAM
OPERA	software package for use on IBM compatible PCs only (386 or greater with 640k of RAM)
PERK	software package for use on IBM compatible PCs only
PLANTRAC-OUTLOOK	software package for use on IBM compatible PCs only (486 or greater with 4Mb of RAM and Windows 3.1)
PREDICT	software package for use on IBM compatible PCs only (386 or greater with 2Mb of RAM, Windows or DOS)
Project Risk for Windows	MS Windows
REMIS	software package for use on IBM compatible PCs only (386 or

	greater with 4/8Mb of RAM, Windows 3.1)
RISK 7000	software package for use on IBM compatible PCs, VMS or UNIX.
Risk Management Database	software package for use on IBM compatible PCs only (386 or greater with 8Mb of RAM) that utilises dBASEIV and interfaces with Wordperfect and Word.
RISK+	software package for use on IBM compatible PCs only (386 or greater with 8Mb of RAM, Windows 3.1) as an add in to MS Project v4.0
RiskNet	software package for use on IBM compatible PCs only (286 or greater)

As stated above a more detailed schedule of risk analysis software is available from the Association of Project Managers.

9
Risk assessment and the CDM Regulations

For the purposes of the CDM Regulations the following definitions apply:

- Hazard is the potential of an operation, substance or condition to do harm.
- Risk is the likelihood of that potential being realised.
- Incident is the failure of the process or operation but without an accident occurring and if it is not reportable to the Health and Safety Executive.

What is a risk assessment?
There are at least two areas where risk assessments are carried out unconsciously over a long period. Firstly, we all make assessments many times a day on the likelihood of undesirable consequences arising from our actions. Examples of these include every time we drive a vehicle. We decide whether to wait at the give way sign or pull out into the traffic. In making this judgement we evaluate the likelihood of injury and its severity.

Secondly, where risk assessment is based upon the requirements for any employer under Health and Safety at Work Act 1974 where in many of its sections, it states that steps are to be taken that are reasonably practicable to ensure the safety of the employees. To carry out this objective means that the degree of risk in any particular activity can be balanced against time, cost, trouble and physical difficulty of taking measures to avoid that risk. Therefore, the greater the risk, the more likely it is that it is reasonable to go to great lengths, e.g. expense, trouble, invention, to reduce it. In reverse, if the consequences and extent of the risk are small or unlikely to occur, then great expense, trouble or invention would not be deemed reasonable.

The difference between those assessments and the ones now required by the Management of Health and Safety at Work Regulations 1992 is that the significant results are to be recorded by the employers and communicated to any persons affected by them.

Risk assessment is carried out to enable control measures to be devised so that persons at risk from any particular operation are made aware of the hazards and are aware of what controlling factors have been implemented.

Risk Assessment is the basis of the MHSAWR 1992 that came into operation on 1 January 1993. As previously described the duty is not new. Health risk assessments are required under the Control of Substances Hazardous to Health Regulations (first edition 1988) and hearing risk assessments are required under the Noise at Work Regulations 1989 to name but two.

Management of health and safety at work regulations 1992 – Regulation 3 Risk Assessment

(1)   Every employer shall make a suitable and sufficient assessment of:

    (a)   the risks to the health and safety of his employees to which they are exposed whilst they are at work

    (b)   the risks to the health and safety of persons not in his employment arising out of or in connection with the conduct by him of his undertaking, for the purpose of identifying the measures he needs to take to comply with the requirements and prohibitions imposed upon him by or under the relevant statutory provisions.

(2)   Every self-employed person shall make a suitable and

sufficient assessment of:
(a)   the risks to his own health and safety to which he is
       exposed whilst he is at work
(b)   the risks to the health and safety of persons not in
       his employment arising out of or in connection with
       the conduct by him of his undertaking, for the
       purpose of identifying the measures he needs to take
       to comply with the requirements and prohibitions
       imposed upon him by or under the relevant statutory
       provisions.
(3)   Any assessment such as is referred to in paragraph (1) or
       (2) shall be reviewed by the employer or self-employed
       person who made it if:
(a)   there is reason to suspect that it no longer valid
(b)   there has been a significant change in the
       matters to which it relates, and where as a
       result of any such review changes to an as
       sessment are required, the employer or self-
       employed person concerned shall make them.
(4)   Where the employer employs five or more employees, he
       shall record:
(a)   the significant findings of the assessment
(b)   any group of his employees identified by it as being
       especially at risk.
Every employer and self-employed person has a duty to
assess the risks arising from their work activities and the
effects that they may well have on other persons either at
the place of work or not. The purpose of the assessments is
to identify any measures which may be appropriate to
protect them.

Where the employer employs five or more persons this
assessment must be in a written format. It must be noted that
these requirements are absolute.

Carrying out risk assessments
A risk assessment consists of identifying the hazards present
and making an estimate of the extent of risk involved, taking
into account any precautionary measures already taken.

A risk assessment is often a very simple, everyday task which is often carried out subconsciously. Sometimes when several construction activities are taking place simultaneously or the operations are complex, the assessment may be lengthy and time-consuming to prepare. In instances that lie between these two extremes, professional advice should be sought to identify hazards present.

The questions to be asked when carrying out a construction risk assessment

Within the construction industry a large majority of all building projects are a repeat of something that has been carried out before. Therefore the experienced constructor will probably have worked previously upon something of a similar nature. The terminology of risk assessment can intimidate the practical constructor but because of operations carried out previously the assessment should be comparatively straightforward. The following questions need to be asked and answered when fulfilling and risk assessment.

Who

- is the client?
- is the principal contractor?
- will be employed to carry out the work?
- else will be working within my work vicinity?
- will be in charge of health and safety on site?

What

- is going to be built?
- construction methods will be used?
- materials will be used?
- are the known hazards?
- are the risks arising from the hazards?
- safe system of work is to be used?

Why

•   is it being built?
•   is a risk assessment required?

Where

•   is it to be built?
•   are the best access points?

When

•   will the job commence?
•   will the job be complete?
•   did the employees receive and health and safety training?
•   do the problems usually arise?

How

•   will the work be carried out safely?
•   can it be proved that it is safe?

Once these questions or the majority of them have been answered, specific issues can be evaluated where individual operations require a risk assessment.

Carrying out specific risk assessments

The project to be carried out must be assessed to establish which operations will require a risk assessment.All assessments must be carried out prior to commencement of the works to ensure that all appropriate protective and preventative measures have been undertaken.

Having identified the hazards, the risks can then be assessed. The risks should reflect the likelihood of harm occurring and the degree of its severity.

Likelihood x Severity = Risk

High	=	Causing death or major injury.
Medium	=	An injury from which the person will be ab sent of work for a period of time.
Low	=	A minor injury which would allow injured person to resume work after treatment.

Where the risk rating is perceived as a high risk ,then the risk must be reduced to as low a level as practicable. Where control measures are introduced in order to reduce the risk, it must be ensured that they are suitable and sufficient. Such measures must be issued to all employees concerned and any other persons affected. Any health and safety training requirements must be carried out prior to commencing the works. It is imperative that all risk assessments are maintained on site for the duration of the contract.

Assessments must be reviewed following an accident, incident or dangerous occurrence or where there has been a significant change in the activity since the previous assessment. It is important to monitor the process on-site with regards to risk assessments because any deficiencies must be rectified immediately.

Protective measures

If an operation is deemed to be a high risk activity then protective measures have to be introduced. When deciding on these protective measure the following principles should be applied.

- Wherever possible the risk should be avoided altogether, i.e. by looking at the operation from a different view.

  e.g   If work was required to a high ceiling would it be better to erect a fully boarded birdcage scaffold rather than carry out the works from mobile scaffold towers.

- It is better to combat the risk at source rather than to provide protective measures. Therefore it is better to prevent a fire from starting than to use fire extinguishers to put the fire out.

- If the above cannot be carried out then the task should be adapted to the worker rather than attempting to adapt the worker to the task.

  e.g.   Redesign storage areas where items are not easily accessible or easy-to-handle. The introduction of mechanical aids should reduce the need for individuals having to bend and stretch.

- Many accidents could and should be prevented by introducing appropriate technology to improve working methods.
- To gain most benefit from risk assessments, priority should be given to implementing control measures which will protect the whole work-place and all the persons who work there.
- Overall risk control can only be effective if it is carried out, implemented and understood by all those concerned.
- All workers need to understand what is required of them so it is necessary to provide adequate training with refresher courses to ensure a level of awareness is maintained. All safety control measures introduced must be seen to be the norm and not an add-on.

There are of course traditionally recognised high-risk activities within the construction process, e.g. roofwork, external cladding, structural steel erection which involve working at height. These operations of roofwork, structural steel and external cladding bring their own specific hazards (see attached generic risk assessments) and if control measures are brought in to reduce the risk, then further risk assessments will be required to evaluate their risk rating.

For working at height the control measures could include general scaffolding, mobile elevating work platforms and aluminium mobile scaffold towers. These control measures can introduce further hazards (see further risk clauses) which require evaluating once the assessor is content that the risk is as low as practicable. The system must be communicated to all persons affected so that they understand the process and what has been put in place. This should ensure that the likelihood of an accident or dangerous occurrence is as low as possible.

Once set up on site the operation must be monitored and any fine-tuning must be implemented immediately.

RISK ASSESSMENTS

Operations covered by this assessment: Erection of steel structures.
Hazards: Inadequately trained operatives. Inadequate or no method statement. Inadequate lifting plant and equipment. Inadequate P.P.E. Not using P.P.E. Equipment in poor condition. Unsafe access and egress. Inadequate working platforms/areas. Weather conditions.
Actions already taken to reduce the risk: Competent specialist contractor appointed. Method Statement provided and approved. All plant and equipment adequate. All inspection and test certificates provided. All necessary P.P.E. provided and worn. Safe access and egress provided. Adequate working areas provided. Adequate protection for third parties. System agreed for work stop due to weather conditions.
Assessment of residual risk: Low/medium/high
Further action required: Agree modifications to Method Statement.

Signed: Position:	Date:

Figure 58. *Risk assessment*

RISK ASSESSMENTS

Operations covered by this assessment: Roofing.	
Hazards: Inadequately trained operatives. Inadequate or no method statement/system of work. Unsafe access and egress. Unsafe storage on roof. Inadequate working platforms/areas. Weather conditions. Inadequate P.P.E.	
Actions already taken to reduce the risk: Competent specialised company appointed. Method Statement provided and approved. Adequate access and egress provided. Safe working platform provided. Roof ladders provided. Necessary P.P.E. provided and worn. Roof ladders provided. Necessary P.P.E. provided and worn. Roof openings protected. Adequate protection provided for third parties. Agreed procedure for work to cease due to adverse weather conditions. Adequate supervision provided.	
Assessment of residual risk: Low/medium/high	
Further action required: Ensure inspection takes place on scaffold structures and F91 completed. Agree any changes in Method Statement.	
Signed: **Position:**	**Date:**

Figure 59. *Risk assessment*

RISK ASSESSMENTS

Operations covered by this assessment: Cladding of buildings.
Hazards: Inadequately trained operatives. Inadequate / no method statement/system of work. Unsafe access/ egress. Inadequate working platforms/working areas. Adverse weather conditions. Inadequate P.P.E.
Actions already taken to reduce the risk: Competent specialist contractor appointed. Method Statement provided and approved. All plant and egress provided. All necessary P.P.E. provided and worn. Adequate protection provided for third parties. All scaffolding erected correctly by competent operatives and inspected. Agreement on when work ceases due to adverse weather conditions. Trained operatives provided. Adequate supervision provided.
Assessment of residual risk: Low/medium/high
Further action required: Ensure scaffolding inspected as required and entry made into F91. Agree modifications to Method Statement.

Signed:	
Position:	Date:

Figure 60. *Risk assessment*

RISK ASSESSMENTS

Operations covered by this assessment: Scaffolding (general). Hazards: Inadequate working platforms. Inadequately trained operatives. Incorrect design. Incorrect erection. Unauthorised interference/modification. Inadequate materials. Inadequate access and egress. O/H and U/G services. Incorrect handling of materials. Ground conditions. Third parties.
Actions already taken to reduce the risk: Specialised company appointed to design and erect. Method Statement provided and approved. All operatives adequately trained. Instructions given to prevent unauthorised interference. Adequate supplies of material available. Safe access, egress and working platforms provided. All services identified. Ground condition stable. Third party protection provided.
Assessment of residual risk: Low/medium/high
Further action required: Weekly inspection and F91 to be completed and maintained.

Signed:	
Position:	Date:

Figure 61. *Risk assessment*

RISK ASSESSMENTS

Operations covered by this assessment: Access equipment (MEWP).	
Hazards: Untrained operatives. Working on/near highways. Overhead services. Falls of material / persons. Overturning. Defective equipment. Adverse weather conditions. Actions already taken to reduce the risk: Operatives to receive adequate training. Clearly cone off working area following sufficient room. Clearly identify services and erect barriers. Safety harness to be used. Do not allow build up of loose materials / tools on platform. Check ground conditions, don't operate on slopes, check swl. Don't travel with platform raised on unsuitable ground. Inspect equipment. Check weather conditions.	
Assessment of residual risk: Low/medium/high	
Further action required: Ensure F91 is filled in, NB Part 1 Section C can be adapted to suit.	
Signed:	
Position:	**Date:**

Figure 62. *Risk assessment*

RISK ASSESSMENTS

Operations covered by this assessment:
Scaffold towers.

Hazards:

Erected above designed height.

Untrained erectors.

Incorrect materials.

Inadequate access, egress, working platforms.

Ground condition.

Traffic adjacent.

O/H and U/G services.

Third parties.

Actions already taken to reduce the risk:

Scaffold not above 3 : 1 in height unless adequately tied.

Trained erectors.

Compatible materials correctly erected.

Ground condition stable.

O/H and U/G services identified.

Adequate protection provided for third parties.

Assessment of residual risk: Low/medium/high

Further action required:

Weekly inspection of F91 to be completed.

Signed:	
Position:	**Date:**

Figure 63. *Risk assessment*

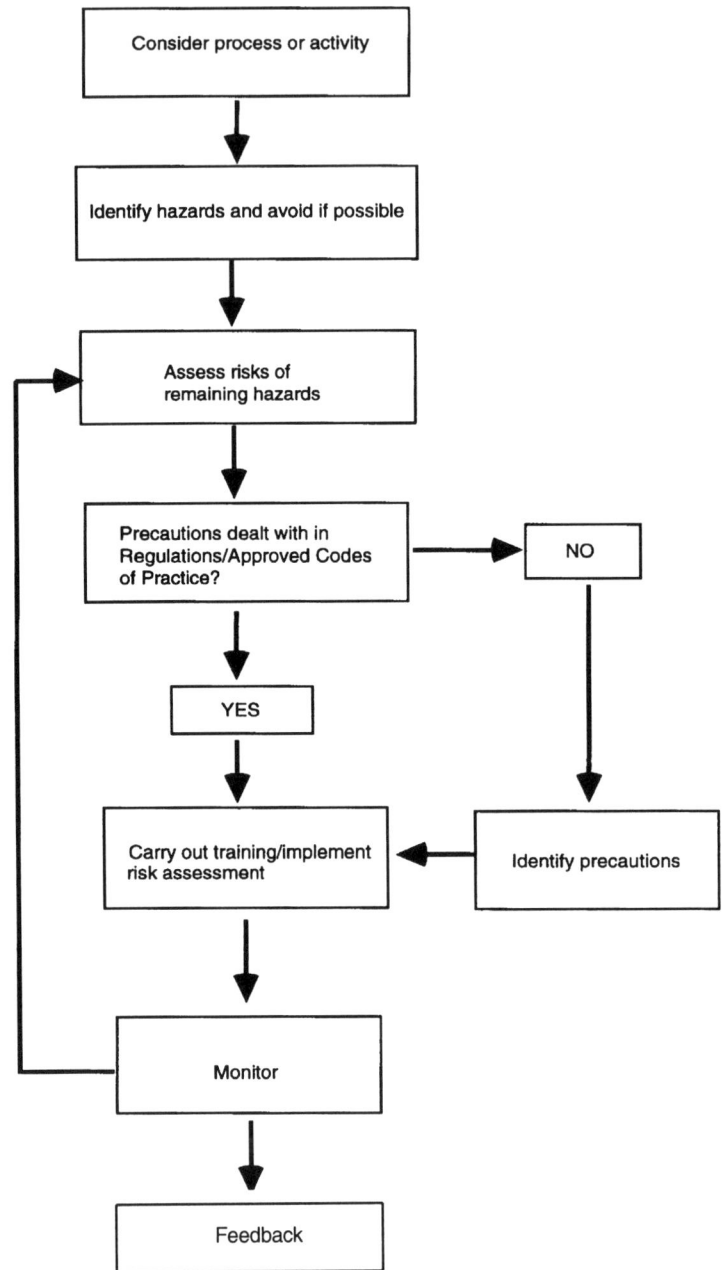

Figure 64. *Risk identification*

The designer and risk assessment

The Health and Safety at Work 1974 Act (HSWA) imposed a duty on all who prepare a design to ensure, so far as is reasonably practicable, that articles and substances are, by design and construction, safe and without risks to health, when being used, set, cleaned or maintained by persons at work.

The Management of Health and Safety at Work Regulations 1992 (MHSWR) reaffirmed this duty upon designers. The Construction (Design and Management) Regulations 1994 (CDM) imposed clear duties upon a designer to carry out risk assessment procedures on a design. The CDM Regulations define a designer as *any person who carries on a trade, business or other undertaking in connection with which he*:

- prepares a design, or
- arranges for any person under his control (including where he is an employer, any employee of his) to prepare a design.

Furthermore, the Regulations defines design in relation *to any structure to include drawings, design details, specification and bills of quantities (including specification of articles or substances) in relation to the structure.*

It therefore seems appropriate to consider the duties of a designer within the CDM Regulations. However, the designer must appreciate that his responsibilities under the HSWA and MHSWA remain, irrespective of whether the design being prepared comes under the scope of the CDM Regulations and that Regulation 13 - Requirements on Designers is a stand-alone regulation.

Designers' requirements for risk assessment

Designers' duties and the CDM Regulations

Under the CDM Regulations designers have a duty to ensure that their design pays adequate regard to health and safety – foreseeable risks should be avoided. If it is not reasonably practicable to avoid them, they should be reduced and controlled.

The term *designer* has a very wide meaning in the CDM Regulations. It includes the following:

- architects and engineers contributing to, or having overall responsibility for, the design
- surveyors specifying articles or substances or drawing up specifications for remedial works
- contractors carrying out design work as part of a design and build project
- anyone with authority to specify, or alter the specification of designs to be used for the structure
- temporary works engineers designing formwork and falsework
- interior designers, shopfitters and landscape architects
- building services engineers designing details of fixed plant.

What is meant by control of risk?

Measures which protect everyone should be given priority over those which protect only an individual. Often, only the contractor can select the right measure, e.g. choosing mobile elevating working platforms in preference to ladder access. However, if the designer is in a position to influence control methods, these principles should be adopted, e.g. fixed rails on a maintenance walkway rather than relying on safety harnesses.

Designers' risk assessments

At the outset designers should aim to:

- identify the significant health and safety hazards likely to be associated with the design and how it may be constructed, maintained and dismantled;
- consider the risk from those hazards which arise as a result of the design being incorporated into the project under consideration
- if possible, alter the design to avoid the risk or where this is not reasonably practicable, follow the remainder of the hierarchy of risk control.

Hierarchy of control

Designers should be aware of the hierarchy of risk control which underlies the modern approach to health and safety management:

- it is best to prevent the hazard and alter the design to avoid the risk.
- if this is not reasonably practicable, the risk should be combated at source (e.g. ensure the design details of items to be lifted include lifting attachments)
- failing this, priority should be given to controls that will protect all workers (e.g. arrange the design to allow the early installation of stairways into the new structure, allow for a one-way system for delivery and spoil-removal vehicles etc.)
- personal protective equipment should only be considered as a last resort when all other control measures have been implemented.

How can designers reduce the risks?

Health and safety needs to be considered at the same time as other design issues. The safest option is not always practicable because of other considerations. The aim should be to select a design option that entails fewer foreseeable risks, within the limits of what is reasonably practicable. The designer should look for ways of reducing and controlling the risks that remain. For example:

- piecemeal construction required at high level may be designed to be prefabricated at ground level, therefore reducing the risk of serious falls
- non-toxic chemicals of comparable performance may be used in many situations, but if this is not possible, potentially toxic chemicals can be specified to be supplied in a diluted form
- incorporation of mechanical aids to help with manual handling and lifting operations
- ample space can be allowed for the maintenance and

replacement of equipment to reduce the likelihood of back strain and other injury from manual handling and lifting operations.

How far are designers expected to go in reducing risk?

The duties on designers when considering health and safety in their design work are qualified by:

(a)   what is reasonable for a designer to do at the time the design is prepared

(b)   by what is reasonably practicable.

In determining (b), the risk to health and safety produced by a feature of the design has to be weighed against the cost of excluding that feature by following the hierarchy of risk control.

The cost is counted not just in financial terms but also in those of fitness-for-purpose, aesthetics, buildability or environmental impact. The overall design process does not need to be dominated by a concern to avoid all risks during the construction phase and subsequent maintenance. By applying these principles it may be possible to make decisions at the design stage which will avoid or reduce risks during construction work. In many cases, the large number of design considerations will allow a number of equally valid design solutions. What is important is that the approach to solving design problems involves a proper exercise of judgement which takes account of health and safety issues.

If, after consideration, it is not reasonably practicable to avoid the risks to health and safety due to buildability, aesthetic or other factors, sufficient information will need to be provided about the risks which remain. This information needs to be included with the design to alert others to the risks which they cannot reasonably be expected to know about.

Attached are examples of designers risk assessments which can be used to show how the hierarchy of control has been carried out and what residual risks still remain that will require the construction contractor to establish their own risk assessments.

Project:		Designer:	
Stage of Work:			
Date:			
Ref.	Hazard	Action	Inform

Figure 65. *Risk identification*

Design Hazard Inventory

Hazard	Persons at risk Site Others	Eliminate by Design	Substitute	Remaining Hazard	Solutions considered not reasonably practicable	File Note Number

Figure 66. *Design hazards inventory*

Hazard reference number	Hazard	Brief Risk Description	Risk rating High Medium Low	Avoidance Control Measure

Figure 67. *Hazard log*

Sample forms

Risk Log			Status	*Confidential*	
Project:	**Refurbishment of existing adult acute ward at DGH**			Date	17/3/96
Sheet	1	of	2	Completed By	AH

Ref	Description	Likelihood	Potential Impact	Response
Site				
1	Access restrictions	L	L	Manage
2	Additional areas of major repairs discovered during execution of works	M	H	Detailed survey; assess & quantify
3	Failure to meet August deadline	M	M	Contract Docs
4	Security	M	L	Contract Docs
5	Amendments to existing lighting	M	M	Investigation req'd
6	Existing drainage alterations	L	M	ditto
Client				
1	Delay in approval process	L	H	Manage
2	Requirement for earlier completion date	L	H	Manage; assess & quantify
3	Reduction / delays in funding	L	H	ditto
3rd Party				
1	Building regulations	L	M	ditto
2	Hospital staff objections	L	L	ditto
3	Environmental Issues	L	L	ditto

Project:	**Refurbishment of existing adult acute ward at DGH**		Date	17/3/96	
Sheet	2 of 2		Completed By		

Ref	Description	Likelihood	Potential Impact	Response
Design				
1	Ability to achieve required 25 year design life	M	H	detail in design brief; assess & quantify
2	Flexibility of design	M	M	Design brief
3	Arrangements for decanting	L	L	Manage (PM to detail)
4	Estimating errors	L	M	Assess & Quantify
− 5	Delays in tendering	M	H	Assess & Quantify
Contractor				
1	Contractor's ability to complete the project within the defined timescales	L	M	Penalty clause in contract
2	Programme	M	H	Assess & quantify
3	Variations	M	M	Assess & quantify
4	Disputes & claims	L	M	Assess & quantify
5	Insolvency	L	M	Assess & quantify
6	Weather	L	M	Assess & quantify
Other				
1	Market conditions: exceptional increases in tender prices	L	M	Assess & quantify

Project	**Refurbishment of adult acute ward at DGH**
Risk Class	Risk Reference _____ S.007
Description of Risk	*requirement for permanent rather than temporary*
protection to plant room due to increased traffic	
Risk Owner *Client*	Responsibility *Design Team*

Proposed Response / Mitigation _____

detailed investigation required by design team to review alternative solutions

Risk Trigger Event _____

Secondary Risks	*yes*	Description	*likely to cause delay in*
			approval process
Reference of			
secondary risk	Cl.001		
Reference of			
secondary risk			

Detailed assessment required	*required/*	By	_____
or undertaken	~~undertaken~~*		
		Filed	_____
Likely Risk Exposure	~~acceptable/insignificant~~/significant/~~critical/unacceptable~~*		

Signed _____   Verified _____
For _____   For _____

INTERIM RISK REPORT	Nr 2 (at option appraisal stage)	
Project:	Proposed alterations to Adult Acute ward at DGH	
Base Estimate	1,250,000	Revised Base Estimate	1,275,000
Risk Allowance	267,000	Revised Risk Allowance	282,500
Expected Value (Budget)	1,517,000	Anticipated Outturn Costs	1,557,500
		Exceeds Budget By:	40,500
Target Completion Date	25/3/97	Proposed Completion	1/4/97
Risk Allowance	12 Weeks	Approved Extensions	4 Weeks
Expected Completion	17/6/97	Risk Allowance	5 Weeks
		Anticipated Completion	3/6/97

Summary of un-resolved risks	Average Allowance	Maximum Allowance*
Access restrictions	27,500	
Alterations to drainage	20,000	
Delay in approval process	15,000	
Decanting issues	175,000	
New Risks		
Permanent protection required to plant room 3	45,000	

* For max allowance see risk analysis

Risk Allowance   **282,500**

Project	**ADULT ACUTE WARD**		

Risk Reference. **S.007**   Date Identified   **21/3/96**

Brief Description of Risk   *requirement for permanent protection to existing plant room*

Identified By   *Design Team (see meeting note 6.11 dated 21/3/96)*

Likelihood of occurence   *low/medium/high**   25%

Potential Impact   ~~*low*~~*/medium/~~high~~**   Impact area   *time/cost/~~quality~~**

Risk Exposure   ~~*acceptable/insignificant*~~*/significant/~~critical/unacceptable~~**

Initial Response / mitigation   ~~*ignore*~~*/manage/design/~~share/transfer~~**

Brief Assessment of Impact

Likely Time   *4 weeks*   Items effected
Worst Case   *26 weeks*

Likely Cost   *£ 45,000*   Description   *protection*
Worst Case   *£ 300,000*   *new plant room*

Likely effect   **none**
on Quality

Secondary Risks   *yes/~~no~~**   Description   *delay in approvals*

Reference of
secondary risk

Responsibilty   **Client**

Detailed assessment required?   *yes/no**   By

Signed _____   Verified _____
For   _____   For   _____

Ref.	Brief Description	Trigger Event	Inter-Dependencies (Ref to)	Ph.	Effect	Impact		Response/Mitigation Status		Owner	Status	Comments/actions
						ARA	MLRA	Current	Previous			
S.001	Access restrictions	investigation		0.40		27,500	68,750					
S.002	Additional repairs			0.25		50,000	200,000					
S.006	Drainage alterations		S.007	0.25		20,000	80,000					
S.007	Plant room	ditto	S.001	0.25	revised design	45,000	180,000					
CL.001	Delays in approvals		S.007	0.25		15,000	60,000	for all responses see individual assessment forms				
CL.003	Funding problems		S.007	0.25								
T.001	Building regulations		S.007	0.10								
T.003	Environmental issues			0.10								
D.002	Flexibility		S.002	0.10		175,000	350,000					
D.003	Decanting			0.50								
CO.001	Contractor's ability			0.10								
O.001	Market conditions			0.01								

T = Third Party Risks
S = Site Risks
Cl = Client Risks
D = Design Team Risks
Co = Contratcors Risks
O = Other Risks

* Ref to is the reference to the dependant risk(s)

T = Time
C = Cost
Q = Quality

A = Assessed
M = Managed out
D = Designed out
S = Shared / Transferred
I = Ignored

I = Initial/intuitive
C = Considered
F = Final

Risk Analysis Schedule -
Option Refurbushment

Base Cost = 17,819,000

Risk	Average Risk Brought Forward	Spread Squared Brought Forward
Third Party Risks	225,000	1.12E+12
Site Risks	222,500	9.43E+10
Client Risks	350,000	4.10E+11
Design Team Risks	90,000	1.25E+12
Contractor Risks	275,000	6.13E+11
Other Risks	1,791,900	4.16E+12
	2,954,400	7.64E+12

Base Cost	17,819,000	
Average Risk allowance	2,954,400	
Average Risk Estimate	**£ 20,773,400**	

Average Risk Estimate	20,773,400	
Square root of spread squared	2,763,473	
Max. Likely Risk Estimate	**£ 23,536,873**	

Third Party Risks	Fixed / Variable	Probability %	Average Risk Allowance	Maximum Risk	Spread	Spread Squared
Changes in building regulations	Variable		100,000	300,000	200,000	4.00E+10
Additional escape stairs	Fixed	50	45,000	90,000	45,000	2.03E+09
Fire regulations	Variable		25,000	75,000	50,000	2.50E+09
Health & safety executive	Variable		25,000	75,000	50,000	2.50E+09
Additional costs associated with statutory authorities	Variable		7,500	176,400	168,900	2.85E+10
Lack of insurance cover	Variable		7,500	1,000,000	992,500	9.85E+11
Vandalism / fire	Variable		15,000	250,000	235,000	5.52E+10
		To Summary	225,000		To Summary	1.12E+12

Site Risks	Fixed / Variable	Probability %	Average Risk Allowance	Maximum Risk	Spread	Spread Squared
Ransom strips	Fixed	25	25,000	100,000	75,000	5.63E+09
Contamination / Asbestos*	Variable		25,000	50,000	25,000	6.25E+08
Listed buildings	Variable		10,000	50,000	40,000	1.60E+09
Site security	Fixed	25	62,500	250,000	187,500	3.52E+10
Changes in water table	Variable					
Flooding	Variable					
Abnormal ground conditions	Variable		25,000	50,000	25,000	6.25E+08
Site availability	Fixed	25	75,000	300,000	225,000	5.06E+10
		To Summary	222,500		To Summary	9.43E+10

Client Risks	Fixed / Variable	Probability %	Average Risk Allowance	Maximum Risk	Spread	Spread Squared
Changes in base directive	Variable		250,000	750,000	500,000	2.50E+11
Prediction of housing needs (see sensitivity analysis included in overall appraisal)	Variable					
Errors in option appraisal	Variable		100,000	500,000	400,000	1.60E+11
		To Summary	350,000		To Summary	4.10E+11

Design Team Risks	Fixed / Variable	Probability %	Average Risk Allowance	Maximum Risk	Spread	Spread Squared
Consultant insolvency	Variable		50,000	500,000	450,000	2.03E+11
Design team skill	Variable		10,000	100,000		
Design demarcation	Variable		10,000	250,000	240,000	5.76E+10
Negligence	Variable		10,000	100,000	90,000	8.10E+09
Bribery / Corruption / Fraud	Variable		10,000	1,000,000	990,000	9.80E+11
		To Summary	90,000		To Summary	1.25E+12

Contractor's Risk	Fixed / Variable	Probability %	Average Risk Allowance	Maximum Risk	Spread	Spread Squared
Insolvency	Fixed	25	250,000	1,000,000	750,000	5.63E+11
Variations	Variable		25,000	250,000	225,000	5.06E+10
		To Summary	275,000		To Summary	6.13E+11

Other Risks	Fixed / Variable	Probability %	Average Risk Allowance	Maximum Risk	Spread	Spread Squared
Tender price index	Variable		1,781,900	3,563,800	1,781,900	3.18E+12
Deflation of rental levels (see sensitivity analysis)						
Landlord / value problems (see sensitivity analysis)						
Terrorism	Fixed	1	10,000	1,000,000	990,000	9.80E+11
		To Summary	1,791,900		To Summary	4.16E+12

Glossary

Average risk estimate or ARE a project estimate in terms of cost and time which is most likely.

Base estimate an evolving estimate of known factors without any element of contingency or allowances for risk and uncertainty.

Change control a method of ordering, investigating and estimating changes to the agreed project definition.

Classification of risks a process which assists in allocating responsibility for risks and identifying causes.

Decision trees a diagrammatic representation of a set of possible alternative solutions (see Chapter 6).

Expected value most likely outcome in terms of time, cost and quality.

Fixed risk a risk that will either be incurred in full or not at all.

Hazard potential cause of risk, for the purposes of this book risk and hazard are generally not distinguished between.

Impact an estimate of the effect the risk will have.

Insurance paying a premium to cover some or all of the cost of the likely impact.

Likelihood this may be expressed on a scale of 0-1 or 0-100%.

Maximum likely risk estimate an estimate which is not expected to be exceeded although it cannot be regarded as the maximum possible cost (often referred to as the MLRE).

Mitigation action to reduce, eliminate or avoid the impact or likelihood of the risk.

Process the procedures which operate within the risk management strategy.

Project risk manager a person designated responsible for the management of project risks.

Qualitative risk assessment a description of the risk, likelihood of the risk occurring and its impact.

Regime the system governing the overall means of reporting on and responding to identified risks. The regime should be described by the project manager as part of his overall risk management strategy.

Risk the possibility that something will go wrong.

Risk allowance the amount by which the base estimate has to be increased to arrive at the ARE or MLRE figure (i.e. average risk allowance or a maximum risk allowance respectively).

Risk analysis the work involved in quantifying risk.

Risk assessment the formal process of assessing risks and identifying their impact.

Risk filter a means by which risks are prioritised for reporting purposes.

Risk log initial schedule of risks used during identification stage.

Risk management for the purposes of this book this is defined as the continual process of responding to and controlling (i.e. managing) risks throughout the design and construction processes.

Risk management strategy the proposed plan for dealing with the overall management of risk. A separate strategy should be evolved, by the project manager, for each individual project.

Risk register a list or schedule of risks identified as being of significance in terms of possible cost implications.

Risk trigger an event that is identified as being a caused effect on the occurrence of a risk.

Risk value an estimate of the cost of the individual risk.

Uncertainty an event that cannot be precisely predicted; uncertainties ultimately result in risk. For the purpose of this book the two are assumed to be synonymous.

Variable risk a risk relating to circumstances which have a variable outcome.

PUBLIC ALLY
number ONE

Tweeds has been partnering public sector organisations for many years working nationwide with local and central government, their agencies and appointed contractors. Projects successfully shared range from new town development to urban and inner city regeneration schemes – and everything else in between.

Just some of our consultancy services include:

Development Monitoring

We advise on all aspects of planning, preparation and procurement.

Embracing engineering and electrical disciplines, as well as construction, we advise on all aspects of planning, preparation and procurement.

Strategic Planning

We provide expert and down to earth guidance on planning, development and the use of new and existing property assets.

Risk Analysis

Our specialists identify and assess areas of risk and propose courses of remedial action including insurance valuations as necessary.

Value Engineering

We help ensure the best balance between time, cost and quality by reviewing projects and design solutions.

Capital Project Control

We assist in financial planning, risk analysis, development appraisal, value judgement options appraisals and life-time costings.

Acquisition of Funding

Assistance with bid preparation for UK & EC development grants including SRB – even national lottery and millennium investments.

Whatever your construction project, trust Tweeds to field an accomplished multi-disciplined team dedicated to reducing risk, adding value and guaranteeing you perfect peace of mind as you proceed to successful completion on time and budget. And with regional offices spread throughout the country you're also guaranteed the highest levels of local understanding and personal commitment.

Initial consultations are free, so you've everything to gain – nothing to lose – by getting in touch with your nearest regional office.

Covering both UK government and EC grants we identify suitable solutions and help formulate application responding to the requirements of grant-making bodies and negotiating arrangements. From SRB to national lottery and millennium investments our expertise can make a crucial difference.

Tweeds

CHARTERED QUANTITY SURVEYORS
COST ENGINEERS · CONSTRUCTION ECONOMISTS

CHURCHILL HOUSE,
160 NEW BOND STREET,
LONDON W1Y 9PA.
TEL: 0171-629 7127. FAX: 0171-629 8279

MARLBOROUGH HOUSE,
11 ST. JAMES'S ROAD, DUDLEY,
WEST MIDLANDS DY1 1HP.
TEL: 01384 255122. FAX: 01384 254845

CAVERN WALKS,
8 MATHEW STREET, LIVERPOOL L2 6RE.
TEL: 0151-236 4502. FAX: 0151-236 0346

50 LEAZES PARK ROAD,
NEWCASTLE UPON TYNE NE1 4PG.
TEL: 0191-232 4256. FAX: 0191-232 5015

ROYAL LONDON HOUSE,
196 DEANSGATE, MANCHESTER M3 3WF.
TEL: 0161-907 3366. FAX: 0161-907 3367

82/84 HIGH STREET,
MOLD, FLINTSHIRE CH7 1BQ.
TEL: 01352 756161. FAX: 01352 756166

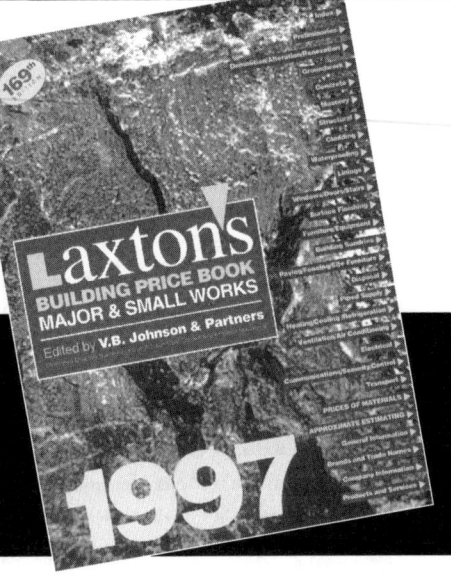
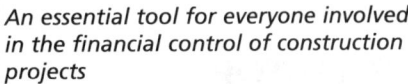

Other Laxton's titles

Laxton's General Specification
Laxton's Mechanical and Electrical Price Book
Laxton's Civil Engineering Price Book
Laxton's Highways Price Book
Laxton's European Building Price Book
Laxton's Industrial Premises Price Book
Laxton's Schools and Colleges Price Book
Laxton's Guide to Term Maintenance Contracts
Laxton's Guide to Budget Estimating
Laxton's Measurement Rules for Contractors' Quantities
Laxton's Guide to Risk Analysis and Management
Laxton's Guide to Single Regeneration Budgets

FOR THE SMALL CONTRACTOR

Laxton's Trade Price Book: Small Works, Repairs and Maintenance
Laxton's Trades Price Book: Electrics
Laxton's Trades Price Book: Plumbing and Heating
Laxton's Trades Price Book: Roofing
Laxton's Trades Price Book: Plastering and Tiling
Laxton's Trades Price Book: Painting and Glazing

SPECIAL OFFER

Customers who buy Laxton's Building Price Book on standing order
are entitled to claim 20% off of any of these other Laxton's titles.
To take out a standing order, check on the availability of the latest
editions of other titles, to order, or to be put on
our mailing lists, use these contact numbers:

BY PHONE: Call our Customer Services Department on **01865 314627**

BY FAX: **01865 314091**

BY E-MAIL: Send to bhuk.orders@bhein.rel.co.uk

BY MAIL: Customer Services Department, Heinemann Publishers Oxford,
PO Box 382, Halley Court, Jordan Hill, Oxford OX2 8RU

WORLD-WIDE WEB SITE URL: http://www.heinemann.co.uk/
Check out our web site for up to the minute
product information, special offers, electronic product demos
and links to other sites of interest to your subject area.

Index

Mode, 56
Monte Carlo Simulation,
   74–77

Net present values, 102–103

Potential impact, 9
Previous experience, 23
Probabilistic analysis, 60–
   65, 107
Probability, 59
Project brief, 7–8
Project management, 12,
   14–16

Qualitative assessment, 37–
   40, 145
Quantitative assessment, 51–
   77
Quantitative analysis, 51–77
Questionnaires, 37–35

Reporting procedures, 79–92
Residual or retained risk,
   49–50
Response, 16
Responsibility, 9
Resultant risks, 14
Risk allocation,9
Risk analysis, 2–10, 146
Risk assessment19, 51–77,
   111–145, 146
Risk avoidance, 46–47
Risk control, 126
Risk efficiency, 4–5
Risk exposure, 42
Risk identification, 18–19,
   23–40
Risk identification form,
   124, 129
Risk log, 131
Risk manager, 14–16

Risk management process,
   2–10, 13–14, 17–22
Risk management strategy,
   11–16, 146
Risk questionnaire, 33–35
Risk reduction, 47–48
Risk register, 43, 146
Risk response, 20–22, 40,
   45–50
Risk retention, 49–50
Risk transfer, 48–49
Risk trigger, 39, 146
RM regime, 78–93

Sample forms, 132–144
Sensitivity analysis, 72–73,
   104–106
Sensitivity factor, 72–73
Severity, 115–116
Simple assessment, 60
Software, 108–110
Sources of risk, 24
Statistics, 55–59
Stochastic dominance, 55

The RM process, 13–16, 17–
   22
Trigger events, 39, 146

Uncertainty, 146
Uncontrollable risks, 24–25

Variable risk, 146